ERDKUNDLICHES WISSEN

SCHRIFTENFOLGE FÜR FORSCHUNG UND PRAXIS

HERAUSGEGEBEN VON EMIL MEYNEN UND ERNST PLEWE

HEFT 40

GEOGRAPHISCHE ZEITSCHRIFT · BEIHEFTE

FRANZ STEINER VERLAG GMBH · WIESBADEN

1975

THE AMAZON RIVER OF BRAZIL

BY

HILGARD O'REILLY STERNBERG

WITH 32 FIGURES AND 2 TABLES

FRANZ STEINER VERLAG GMBH · WIESBADEN
1975

Zuschriften, die die Schriftenreihe „Erdkundliches Wissen“ betreffen, erbeten an:
Prof. Dr. E. Meynen, 532 Bad Godesberg-Mehlem, Langenbergweg 82
oder
Prof. Dr. E. Plewe, 69 Heidelberg, Roonstraße 16

ISBN 3-515-02075-6

 Gesamtherstellung: Limburger Vereinsdruckerei GmbH, 6250 Limburg/Lahn

Printed in Germany

RICHARD JOEL RUSSELL

IN MEMORIAM

CONTENTS

FIGURES

This assemblage of information and reflections on the Amazon River of Brazil has no pretentions of being a structured and exhaustive study. Its focus lies within a domain of generally low equatorial and tropical lands. A major part of them is forested and known as the Hylaea. In Brazil, the expression Amazônia at times is used as a synonym of Hylaea, at times in a broader sense, even to define an operational region bounded in part by meridians and parallels ("Amazonia Legal"). Although the lowland "Amazonian look" (Spruce, 1908) of the riverscape continues well into Peru, the *triplex confinium* of Brazil, Colombia and Peru serves as the break-off point for the present review.

The name Amazonas applied to the river can have two different connotations. One, to designate the entire trunk stream, from mouth to source--and there has been quite some dispute as to which affluent should be considered the headstream. The other, to designate a specific stretch of the river. Thus, according to Brazilian usage, the designation Amazonas prevails from the mouth of the Rio Negro downriver. Above the confluence, as far as the international border, the master stream is known as the Solimões. In this *aperçu* "Amazon" is used to designate the Solimões-Amazonas; "Amazonas", to specify the section below the embouchure of the Negro.

Most quantitative data presented are no more than provisional approximations.

The study is, in part, a by-product of several seasons in Amazônia, beginning in 1948, when the writer's research was sponsored by the Federal University at Rio de Janeiro. Later, at the University of California, Berkeley, field work has been made possible by grants from the Campus' Center for Latin American Studies and Committee on Research. Acknowledgement for cartographic work is due to Mrs. A. D. Morgan, of the Department of Geography.

THE MAIN RIVER

THE RIVER AND THE SEA

The *Mar Dulce* or Sweet Sea, which the explorers under Vicente Yáñez Pinzón found themselves navigating in 1500, as they cruised the coast of South America, generally is identified with the Amazon River (*Fig. 1*). The description of a body of water so fresh that a ship's casks could be filled from it at sea, so powerful that it pushed back the ocean itself, was one to excite the imagination. The legend persisted for centuries, although the facts hardly require such embellishment: according to the most recent estimates (USGS, 1972), the average discharge of the Amazon is of the order of 160,000 m^3 per second, *i. e.* more than four times the mean flow of the Congo, the second largest river on earth, and about ten times that of the Mississippi (*Fig. 2*). The Amazon's contribution to the Atlantic accounts for approximately 15 per cent of all the fresh water passed into the world's oceans. Entrained by the northwesterly Guiana Current, the Amazon outflow, almost alone, lowers the salinity of the sea in a measurable degree over an area estimated to exceed at times two and a half million km^2 (Ryther, *et al.*, 1967).

The magnitude and ultimate destination of the Amazon's sedimentary load have long been a matter of speculation. Although actual measurements made in the 1960's (*Fig. 3*) revealed the Amazon's discharge to be far greater than that indicated by computations based on rainfall-runoff relationships, the concentration of suspended solids seems to have been overrated in the past. Even so, the total volume of particulate matter cast into the sea exceeds 1 1/3 million tons per day (Oltman, 1973).

Observations concerning transport of this debris by ocean currents and the characterization of the coast "from the Amazons to the Oroonoko" as having been for ages the "resting-place for the washings of great part of South America" (Lochead, 1798) were given wide currency by Charles Lyell in his *Principles of Geology* (1832). The tearing down of the land and the deposition of the scourings on faraway shores are themes that can inspire eloquence. Hartt (1896) thought of the 100-armed Briareus of Greek mythology, as he described the Amazon, grasping for its tribute in eroded soil by means of far-stretching affluents. Euclydes da Cunha (1909) is vehement in his comments on the destination of the tribute thus exacted. A Brazilian, he contends, who lands in the Guianas or even on the coast of Georgia or the Carolinas, though an alien, really sets foot on native soil. The Amazon appears to Cunha as a strange adversary that night and day saps its own land (by which he meant Brazil) and "builds its real delta in remote parts of the other hemisphere." His reproach: "the river that more than any other provokes our patriotic lyricism is the least Brazilian of rivers."

Insofar as ongoing demolition of the landmass of Brazil is concerned, the Amazon does not quite deserve Cunha's censure, as will be seen later. Nor does the detritus brought down by the river bear upon the rate of accession of the southeastern United States. Nevertheless, some of it appears to have been transported by turbidity currents

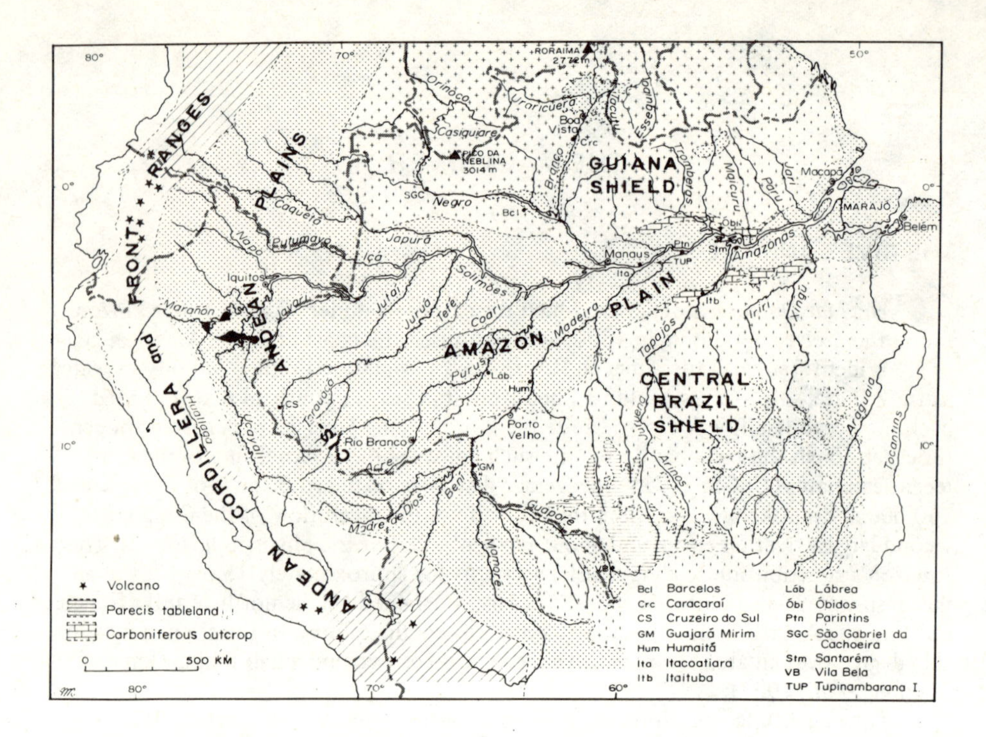

Fig. 1. The Amazon Basin. The superficial deposits of the Amazon Plain, with its uplands and floodplains, overlie a deep, sediment-filled structural basin or trough. Dipping toward the axis of the basin, Paleozoic and Mesozoic strata are buried under younger accumulations, except along the edges of the plain in the eastern portion of the valley (only outcrops of Carboniferous age are indicated).

The Cis-Andean Plains decline eastward and are formed essentially by a mantle of debris (including some volcanic materials) from the cordilleras and front ranges. In the west, this cover of Tertiary and Quaternary age blankets older sedimentaries deformed by the Andean orogeny. In the east, it merges with the Amazon Plain or laps over the Guiana and Central Brazil shields, *e. g.* in the Beni River Plains of Bolivia. The sketch, being very generalized, does not show such features as the *serra* (exceeding 600 m elevation) on the Peruvian boundary of Acre, where one of the subandean folds surfaces through the overlying sedimentaries and is drained by left-bank tributaries of the Juruá, upstream from Cruzeiro do Sul (Bischoff, 1963a).

The bevelled igneous and metamorphic rocks of the Shields are overlain here and there by subhorizontal sedimentary strata, as in the almost circular lowland of the upper Xingu River or the sandstone Parecis tableland.

Not represented, except for the Casiquiare, are the several points where the Amazon drainage communicates with or allows easy portages to neighboring basins.

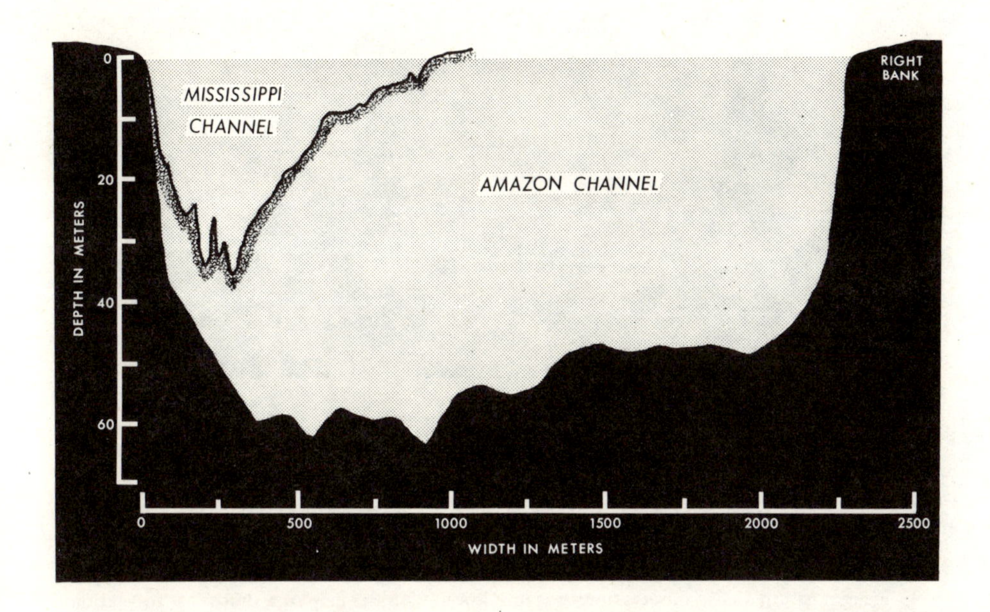

Fig. 2. Comparison of the channels of the Amazon at the Óbidos narrows and the Mississippi at Vicksburg (on July 16, 1963 and April 6, 1964, respectively). Plotted by R. E. Oltmann. Six measuring runs made at Óbidos on June 5, 1972 averaged 126,000 m^2 for the cross-section and 232,000 m^3/sec for the river flow--the largest discharge ever recorded on earth; an additional by-pass flow through a lake south of Óbidos was estimated at 23,000 m^3/sec, giving a total of 255,000 m^3/sec (Smoot, 1972). The 1963 gaging team recorded several soundings of the order of 100 m in the Amazonas as far upstream as the Rio Madeira (Oltmann *et al.*, 1964), including one in the narrows proper, directly in front of Óbidos. For this reason, measurements were carried out some 2 km downstream, in the rather uniform section represented here, still within a reasonably narrow reach of channel, but with a maximum depth of the order of 60 m. The 1963–64 reconnaissance investigations indicate little variation in cross-sectional area at this point, and suggest that increased discharge is achieved largely by virtue of an acceleration of flow through the narrows (Oltman, 1968).

Fig. 3. Gaging the rate of flow of the Amazon River. The first discharge measurements made in 1963 and continued in 1964 revealed an average flow 1/3 greater than the most liberal estimates. Soundings and velocity measurements on a section about 2 1/2 km wide, just downstream from the "narrows" at Óbidos, were made from a Brazilian corvette, whose exact position during observations was monitored with the help of a shore control station. A shipboard tellurometer gave precise ship-to-shore distances by means of microwaves reflected from the unit seen in photograph; horizontal angles were measured with a conventional theodolite located on the bank. (H. O'R. S.).

considerable distances across the sea bottom, contributing during the last half-million years to the partial filling of a deep oceanic trench, more than 500 km to the northeast of the present river mouth (Shipboard Scientific Party, 1971). Contemporary research also confirms the paramount role of Amazon sediments in the development of the shoreline stretching northward to the Orinoco. They are responsible for the muddy shoals noted by La Condamine (1749), when he sailed from Pará to Cayenne: "two months of navigation by sea, & even by land . . ., without exaggeration, since . . . the helm did not cease to furrow the ooze, there not being sometimes a foot of water at half a league's distance from shore."

The coastwise movement of sediment of Amazonian origin is linked to two major problems that afflict the littoral of the Guianas. One is the difficulty of access to navigable rivers and the blocking of drainage outlets by the choking mud. The other is the erratic way in which accretion alternates with erosion along given sections of the coast, as gigantic ripples of Amazonian sediments travel along the bottom of the brown muddy sea, now shielding, now exposing the shore to wave attack (Reyne, 1961; Hydraulics Laboratory, 1962; Veen, 1970).

THE AMAZON AND RIO PARÁ ESTUARIES

Most of the Amazon waters and the sediments they carry are discharged through a channel, subdivided by low islands, the Canal Norte that forms the northwest perimeter of Marajó Island. With about 50,000 km^2, larger than Switzerland, half selva, half low-lying grasslands that support more than half a million head of cattle, Marajó is bounded on the south and east by the so-called Rio Pará, a sort of cul-de-sac connected to the Amazon proper by a maze of deep narrow channels. These separate Marajó from the mainland and integrate a labyrinthine lowland known as the region of the *furos* (passages) (*Fig. 4*). Although tidal action (the most spectacular manifestation of which is the bore or *pororoca*) reverses the flow in much of this vascular system, the net result is an influx of Amazon waters into the Pará estuary. Because of the smallness of this contribution, the Rio Pará hardly can be considered a mouth of, nor the Tocantins River, which debouches into it, a tributary to the Amazon.

When, on Christmas Day 1615, a small armada weighed anchor at São Luís, Maranhão, and set a course westward for the Amazon, there to establish Portuguese hegemony, the ships, after skirting the shore for 18 days, found their way through a narrow passage and swept into the wide embayment of the Rio Pará. The expeditionaries may have mistaken this sinus for the mouth of the Amazon (see Pereira's 1616 *Relação* . . ., 1889). A blockhouse was built, Fort of the Crèche (Presépio), and in its shelter the town of Our Lady of Bethlehem (Belém) grew, to become today's bustling, if somewhat off-side, portal to the Amazon (*Fig. 5*). In the 18th century, seeking to safeguard the Canal Norte, the Portuguese built one of the most important fortresses of colonial times at Macapá (*Fig. 6*). Yet most ocean-going ships still ply their way up the Amazon through the furos, after putting in at Belém (1970 pop. 565,000). In fact, it is only recently that the northern channel has begun to come into its own, with the shipping of manganese ore from bulk-loading facilities built at Santana, a deep-water port 22 km upstream from Macapá (*Fig. 7*). The more direct route cuts two or three days from the ore ships' round trip to European and United States markets. However, despite detailed hydrographic surveys and the installation of beacons and buoys, the rapidly shifting position of shoals and channels requires that the passage be negotiated with caution.

THE AMAZON PLAIN

The sites of the two old fortifications at Belém and Macapá command a view of the estuarine waters from a bluff five to ten meters high. Both were built on the edge of a terrace, one among many levels of a vast sedimentary plain estimated to occupy some 1.2–1.3 million km^2 in Brazil alone. The area commonly known as the Amazon Plain lies between two rigid blocks of very old, folded and faulted metamorphic and igneous rocks, overlain in places by younger, subhorizontal strata; these blocks, because of their stability, are called the Guiana and Central Brazil Shields. The designation

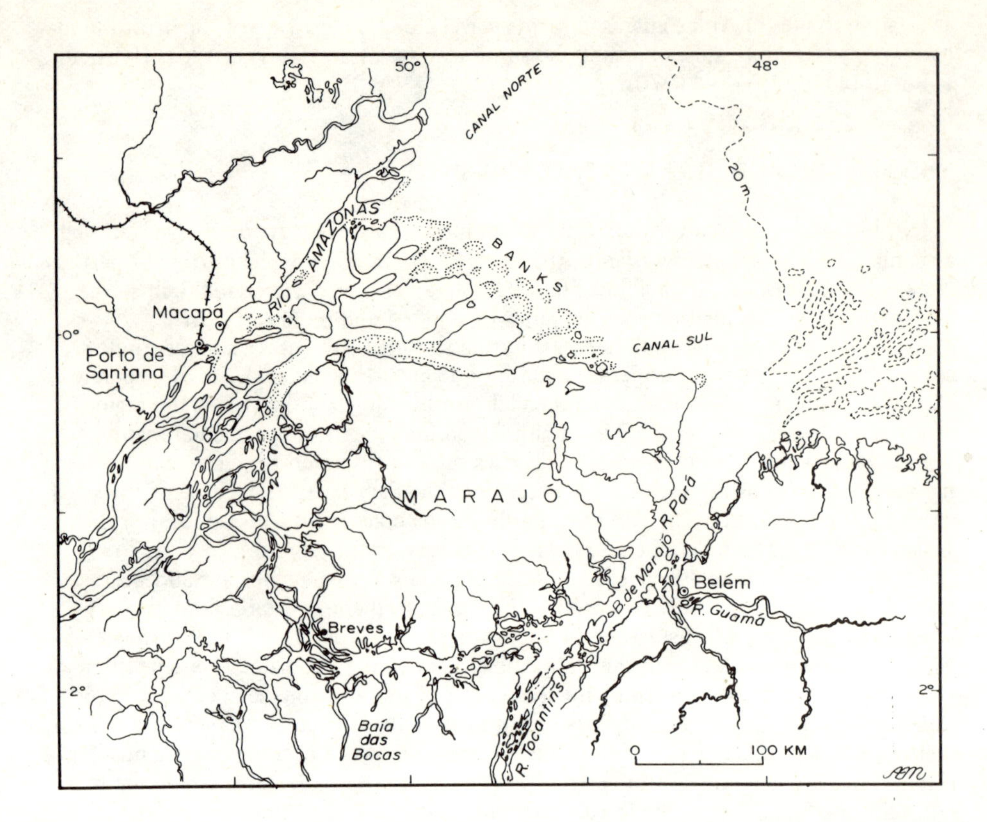

Fig. 4. The Amazon and Rio Pará estuaries. Simplified and reduced from Marinha do Brasil, Diretoria de Hidrografia e Navegação, Chart No. 40, *Brasil Costa Norte, da Baia do Oiapoque ao Rio Parnaíba.* 3rd. ed. 1971, 1:1,159,563. Only the 20 m isobath is shown here (datum: approx. mean syzygial low water). The maze of channels between the Amazonas and the Pará is known as the *"furos* of Breves". Except in the case of ore boats that load at Porto de Santana, it constitutes the preferred route of access to the Amazon: banks and channels at the mouth proper are constantly changing, and mariners are warned to proceed with caution.

Fig. 5. Belém. The traditional façade of the old business district, the "Comércio," out of which a half dozen buildings thrust skyward, gives upon the river and the "Ver-o-Pêso." This small rectangular basin (lower right), animated with the comings and goings of colorful sailboats, was the site of the first "port" of the settlement that grew up around Forte do Presépio. Belém's influence over the immense hinterland served by the Amazon River network permitted the old rubber capital of the world to rally after the keystone of its economy had come tumbling down–in 1970 it placed 7th among Brazil's burgeoning cities. (H. O'R. S.).

Fig. 6 Fort São José do Macapá. Erected during the period 1764–82 on the left bank of the Amazonas to resist possible French advances from Guiana, and as a protection against English and Dutch freebooters. Fort Macapá was the estuarine anchor of a semi-circular chain of river bank fortifications, whose distal end was tied to another monumental bastioned fortress, Príncipe da Beira, on the banks of the Guaporé, where this river flows into the Mamoré. Except for an outpost against the Dutch in the upper Rio Branco, the vast interior defense perimeter of Portuguese Amazônia was deployed essentially against the Spaniards. Fort Macapá dominates more than 10 km of open water from a bluff of vesicular lateritic material, which is retreating under the brunt of wave attack. The stronghold itself is built of plinthite (laterite) blocks. (H. O'R. S.).

Fig. 7. Port Santana. Manganese ore loading facilities on the left bank of the Amazonas, 22 km upstream from Macapá. The deep-water embarcation point on a channel between the mainland and Santana island is equipped with a floating wharf. The washed manganese ore is hauled by diesel-electric locomotives from the open cut mines 194 km away, and loaded into ships with the help of a conveyor system. (H. O'R. S.).

Amazon Plain sometimes is loosely applied to all Hylaean lands, regardless of geology. Actually, the junction between the sedimentary center and the bevelled oldlands can be quite inconspicuous to the untrained eye, especially since it is generally masked by lush vegetation. It does find obvious expression however, in the rapids or falls that roughen the waters of many tributaries at and upstream from the contact.The Amazon Plain, at its narrowest near the mouth of the Tapajós, flares westward to merge with the Cis-Andean Plains that extend in an immense semi-circle from Colombia to Bolivia and in turn are embraced by the great wall of the Cordilleras.

The rocks that crop out in the Amazon Plain represent the topmost layers of a thick sequence that has been accumulating since Paleozoic times in a series of vast, interconnected east-west structural basins. All but completely buried under Tertiary and younger accumulations, the older (mostly Paleozoic) strata surface only in two narrow, roughly parallel belts east of Manaus, where the sedimentaries abut against the stable masses of Guiana and Central Brazil. A part of these belts, rocks of Carboniferous age, are represented on figure 1. The bottom of the pile in places lies more than 5,000 m or, as indicated for a graben at the mouth, 6,000 m below sea level (Ferreira *et al.,* 1971) and the subsidence of the earth's crust that took place as the sediments were laid down, compensated for by uplift elsewhere, has been accompanied by warping and cracking of the basement and overlying materials. These effects are discernible on the surface, because of their influence on drainage patterns: approximately northeast and northwest alignments prevail in many streams or stream segments that run parallel or at right angles to one another. Rectangularity in a region where the underlying strata are essentially horizontal suggests that the fluvial network is exploiting the surface trace of a conjugate system of fractures or faults (Sternberg, 1955; Sternberg and Russell, 1957).

Two basically different groups of landforms obtain in the sedimentary axis of Amazônia: (1) the *terras firmes,* or uplands, of Tertiary and Pleistocene materials, which lie above the highest flood level; and (2) the *várzeas,* or floodplains, of Recent alluvium, which stretch along the trunk stream and some tributary valleys (*Fig. 8*). The várzeas have loomed disproportionately large in the way the Amazon Plain is perceived as an immature land, "an unfinished . . . and contemporary page in the Book of Genesis" (Cunha, 1907). This image, conjured up by the unceasing erosion and deposition that occur in the alluvial strips, is totally inappropriate for the rest of the Plain. The realization that most of the land covered by the Amazon forest is made up of low plateaux, terraces, and, in some parts of the crystalline basement, rounded hills, and that it includes tracts with considerable local relief, was brought home forcefully when work began on the Transamazon highway in 1970.

The Terras Firmes

The existence of multiple topographic levels in the sedimentary strata, which have been entrenched by and now overlook the rivers, is widely recognized. There is little agreement, however, in assigning specific elevations to and in explaining the origin of such planes (Sakamoto, 1957, 1960; Sombroek, 1966; Ab'Saber, 1967; Klammer, 1971).

Dominating all other surfaces that top off the sedimentary fill are mesa-like remnants of a widespread, essentially flat-lying stratum of relatively unconsolidated sands, silts and clays, believed to be of Tertiary age. There are widely divergent viewpoints concerning the environment in which these beds originated (Sakamoto, 1960; Ab'Saber,

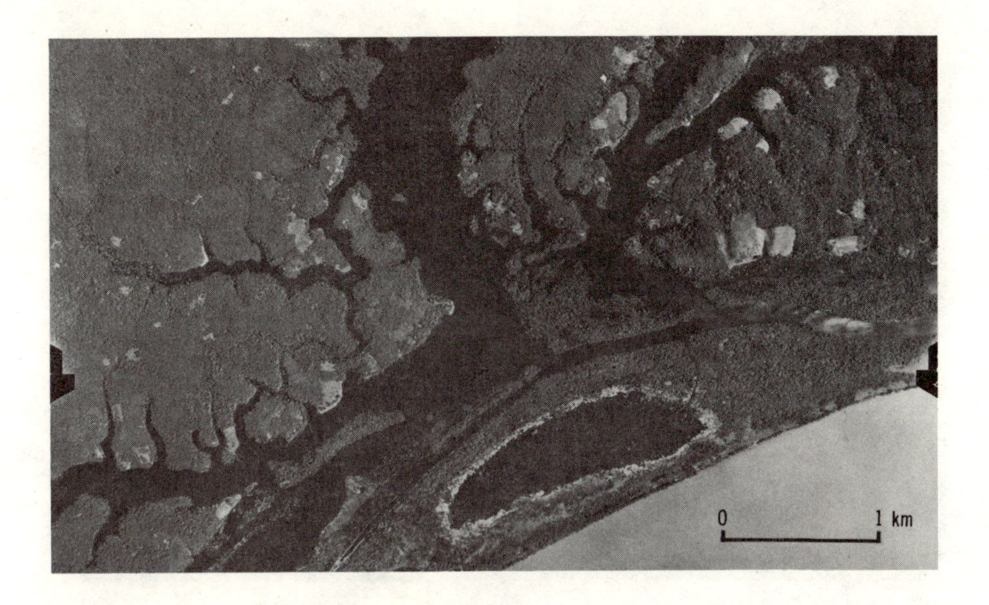

Fig. 8. *Terra firme* and *várzea*. Amazon, about 12 km downstream from Negro. Drainage pattern of terra firme, with more or less straight stream segments at right or oblique angles, is influenced by a fine-grained system of joints. Eluviation along these joints may contribute to give them topographic expression (Bremer, 1974). Recent deposition of Amazon sediments across the arcuate bluff has blocked ria-like valleys, transforming them into terra firme lakes. It also created a várzea lake; silty floodwaters aided by accumulation of plant remains will gradually build up this low area. (Cruzeiro do Sul S. A.)

1967; Mabessone, 1967; Klammer, 1971). Whatever the nature of the depositional episode, it was brought to a close, and fluvial dissection took over. As the Amazon and its tributaries bit into the sediments, portions of these remained standing and constitute today's low tablelands. One such remnant may be observed near Santarém, at the site of the former Ford rubber plantation, where a surveyed profile reaches some 130 m above the low-water level of the Tapajós (Klammer, 1971) and the terrain probably rises to somewhat higher elevations at a distance from the river. This is the type locality for what has been called the Belterra plateau (*Fig. 9*); the fact that apparently the same surface is found elsewhere at considerably different altitudes (*e. g.*, about 200 m on the Jari River, not far from Macapá) can be explained by warping or tilting or differential vertical movements of crustal blocks.

Sea Level: Entrenchment and Alluviation

Throughout the entire Pleistocene and possibly late Tertiary history, the geo-

Fig. 9. The Belterra plateau. The *planalto* south of Santarém is the type locality of widespread, essentially flat-lying sediments, of presumed Tertiary age. The upland declines step-like to the drowned estuary of the Tapajós (far right background), which it dominates by some 130 m (at low water, *cf.* Klammer, 1971). Although the Belterra plateau probably is somewhat more elevated farther away from the river, its general altitude is considerably smaller than that of several other upland flats in Amazônia, believed to be relicts of the same surface, such as a 200 m (approx.) level near the Jari River. Klammer suggests that such differences in elevation are tectonically influenced. (H. O'R. S.).

morphic evolution of the Amazon Plain was achieved largely through a process of alternate cutting and filling, keyed to marine transgressions and regressions; these, in turn were associated with the waxing and waning of distant continental glaciers.

For an admittedly simplistic interpretation, consider the earliest "glacial" stage to affect the landforms of the Plain. With water locked up in the form of ice on middle- and high-latitude lands, sea level fell worldwide: the Amazon and other rivers draining the continents adjusted to this change by deepening their beds, thus inviting a corresponding incision of affluents and subaffluents. With the end of this glacial stage, sea level rose again, backing up the Amazon and making the previously entrenched river into a freshwater gulf or ria, several thousand kilometers long, with ramifications leading up tributary valleys. Sediments contributed by inflowing streams began to fill the drowned valley systems; however, with the onset of the second glacial stage and a new drop in sea level, the rivers cut into and flushed out most of this alluvium. Part of what survived, perhaps slowly elevated by the already mentioned crustal movements, came to stand above the younger alluvial inlay, deposited during the second "interglacial"--

i. e., high sea level--period. The sequence of events described was repeated several times, creating a set of nested terrace levels.

One can begin to appreciate the true complexity of the geomorphological problems involved if it is remembered that, in addition to the eustatic and tectonic controls briefly referred to here, climatic changes *within* Amazônia may also have played a decisive role. It has been hypothesized, for instance, that certain "terrace" levels are, in fact, remnants of surfaces elaborated under far drier conditions than now prevail (Ab'Saber, 1967). Also, in relation to eustatic controls, there is the ongoing revision of Pleistocene chronology, a reassessment of the long-recognized concept of four glacial advances (see, for example, Cooke, 1972).

Since a thorough discussion of Amazonian landform evolution does not lie within our purview, this section may be concluded with the remark that alluviation, in the course of and subsequent to the last rise of sea level, has advanced down the valleys that were cut during the youngest glacial stage, and, by filling the ramified embayment, has produced the present várzea surface. The task, however, has not been fully achieved in the entire river network, and several large water areas remain open in the wide lower courses of a number of tributaries.

Streamload

The fact that some major affluents have been able, others unable to fill the lake-like lower sections of their courses is explained by the varying amounts of solids they carry. The load of Amazonian rivers, as well as many other particularities, such as the character of the riparian and aquatic vegetation or the faunal species that inhabit the waters, traditionally has been associated with the optical behaviour of the streams. This criterion had been widely incorporated in aboriginal river names (*e. g.*, suffix *una* = black; *tinga* = white), and the first Europeans to travel down the Amazon River also were impressed by the specially striking inkiness of one major affluent, which since then has been known as the Rio Negro. The threefold division of Amazonian streams later used by Alfred Russell Wallace (1853) continues to be followed by contemporary students of Amazônia: "white-," "blue-," and "black-water" rivers.

Wallace also recognized the association between optical properties and different source areas. "White" rivers--actually muddy-yellow, due to the sediments they transport--most characteristically issue from the Andean Cordillera and its foreland. Clear-water rivers, perceived by some as blue, by others as green to olive green, rise in the Guiano-Brazilian shield areas. Relatively short rivers that head in the sedimentary uplands of the central Amazon basin also carry very small quantities of sediments and normally belong to this category. The third group, of which the Negro is the prototype, consists of streams whose waters, held in a glass, prove to be tea-colored, but in the river may appear jet black. They are similar to clear-water streams, in that they carry minimal amounts of inorganic sediments; dissimilar, because of their extremely high quota of dissolved humous substances. Following up an initial paper in 1954, Sioli with his collaborators has been carrying out geochemical investigations to explain the origins of black-water streams in areas of bleached sands or podzols (Sioli, Schwabe and Klinge, 1969). The earliest description of this phenomenon may have been that of Lochead (1798), who, "while on a botanical excursion to the Dutch colony of Demerary," observed springs issuing from sparsely vegetated sand hills with inclusions of black vegetable earth. The waters, "notwithstanding the extreme whiteness and purity of the sand from whence they flow, . . . come out . . . of a brownish colour, very much like the water which runs from peat-mosses, and they are certainly tinged by the same cause."

The first basin-wide investigation of the dissolved and suspended load of Amazonian streams (Gibbs, 1965, 1967) has resulted in a quantitative assessment of the contribution to the total discharge into the Atlantic by tributaries, according to the threefold categorization adopted by the author of the study. Of the fifteen rivers researched, two derive their flow and load from the high-mountain environment of the Andes. Nine others originate in and course through lands of low to medium elevation, mainly the old Guiano-Brazilian shields and/or the sedimentary Amazon Plain. The remainder are mixed streams that rise in the environment characteristic of the first group, but also receive an inflow from that of the second (*Table 1*). Some results of the pioneer investigation that concern dissolved salts and suspended solids are summarized in table 2. Figures for suspended solid erosion (based on calculated discharges and measured concentrates) indicate how very little the oldlands of Brazil and Guiana are contributing to the total sediment discharge of the Amazon. The stability of these worn-down lands has permitted a great depth of weathering (*e. g.* in excess of 100 m at one site in Amapá, *cf.* Nagell, 1961). The streams draining them carry little dissolved salts and suspended solids. The very low rates of erosion of particulate matter in catchments like those of the Tapajós and Xingu are consistent with the fact that these rivers have yet to fill in their drowned lower courses with alluvium. By contrast, suspended solids are plentiful in headwaters that cut deep into Andean Cordilleras and front ranges, quarrying fresh material from the slopes and clearing out residual alluvium from the mountain valleys, or pick up debris from the piedmont plains. It has been estimated, in fact, that 82% of the total suspended solids discharged by the Amazon derive from 12% of the total area of the basin, namely the Cordilleran environment (Gibbs, 1967).

At this point, one recalls how the Amazon River was blamed for making free with Brazilian soil. As it turns out, the river system actually is the purveyor of soil--and of nutrients--for Brazil. It broaches the rich storehouse of the youthful Andean mountains and their foreland, areas of complex rock types, unweathered and rich in mineral salts. Sediments derived therefrom have contributed generously to the fill of drowned lowland valleys. The fertility of this allochtonous alluvium is set off in central Amazônia by the general poverty of the surrounding terra firme soils, derived from local parent material, Plio-Pleistocene sedimentaries. Short streams heading in these areas tend to be wanting in mineral salts, acidic (pH around 4.5) and poor in aquatic plant and animal life (*e. g.*, no water snails).

Waters originating in limestones and gypsites of the flanking belts of Carboniferous age (type locality: Itaituba) are more or less neutral and relatively nutrient-rich. If on the one hand their productivity is higher, on the other they offer a favorable milieu for planorbid snails, the intermediate hosts of *Schistosoma mansoni,* as first pointed out by Sioli (1953a, 1953b) in relation to the Fordlândia rubber estate on the Tapajós River. Source areas such as represented by the Carboniferous strips therefore are among those where the danger is greatest that schistosomiasis (bilharzia) might become a major health problem.

Table 1.

Area and Computed Discharge Data for Amazon and Selected Tributary Basins, by Environment

Basins	Area 10^3 km^2	Discharge 10^{12} m^3 / yr.
Amazon at mouth	5,916[a]	6.773[b]
Cordilleran		
Ucayali	406	0.301
Marañon	407	0.343
Plains/Plateaux		
Xingu	540	0.243
Tapajós	500	0.224
Purus	372	0.341
Coari	55.5	0.054
Tefé	24.4	0.025
Juruá	217	0.197
Jutaí	74	0.076
Javari	106	0.116
Negro	755	1.407
Mixed		
Madeira	1,380	0.992
Japurá	289	0.351
Içá	148	0.180
Napo	122	0.145

Sources: Gibbs, 1967, except [a]Brasil, 1967; [b]USGS, 1972.

Table 2.

Erosion Rates for Amazon and Selected Tributary Basins, by Environment

Basins	Dissolved Salts		Suspended Solids	
	$\frac{10^6 g}{km^2/yr}$	$\frac{10^{12} g}{basin/yr}$	$\frac{10^6 g}{km^2/yr}$	$\frac{10^{12} g}{basin/yr}$
Amazon at mouth	36.8	231.8	79.0	498.5
Cordilleran				
Ucayali	152.0	61.4	307.1	124.6
Marañon	92.8	37.8	251.5	102.4
Plains/Plateaux				
Xingu	2.8	1.5	0.9	0.5
Tapajós	3.8	1.9	1.2	0.6
Purus	30.3	11.3	43.2	16.1
Coari	13.2	0.7	2.0	0.11
Tefé	12.4	0.3	2.2	0.05
Juruá	33.2	7.2	49.4	10.7
Jutaí	5.0	0.4	40.8	3.0
Javari	11.5	1.2	68.3	7.2
Negro	10.0	7.5	10.1	7.6
Mixed				
Madeira	42.4	58.5	157.3	156.9
Japurá	10.9	31.6	120.2	34.7
Içá	17.0	2.5	61.9	9.2
Napo	29.1	3.6	184.0	22.4

Source: Gibbs, 1967.

The Várzeas

Granted that the floodplains may represent a fraction of no more than 1–2% of Brazil's share of Amazonian lowlands, these are of continental dimensions. One estimate gives the várzeas a total of 64,400 km^2, an area almost double that of the Netherlands (Camargo, 1958).

Channel Scour and Aggradation. Whereas in the terra firme the rate of morphological change generally is slow, the várzeas' constant evolution is rapid enough to invite attention. The Amazon and its white-water tributaries rework the floodplain sediments, cutting their banks at some places and building new land at others. The main cause of bank caving (*terra caída* = fallen land) is the scouring of the stream bed by macroturbulent eddies, which may be observed directly in their final stages, as they surge to and dissipate at the surface. The vortical action plucks material from the bottom and causes localized deepening of the bed; stability destroyed by oversteepening of the bank is reestablished by slumping (*Fig. 10*).

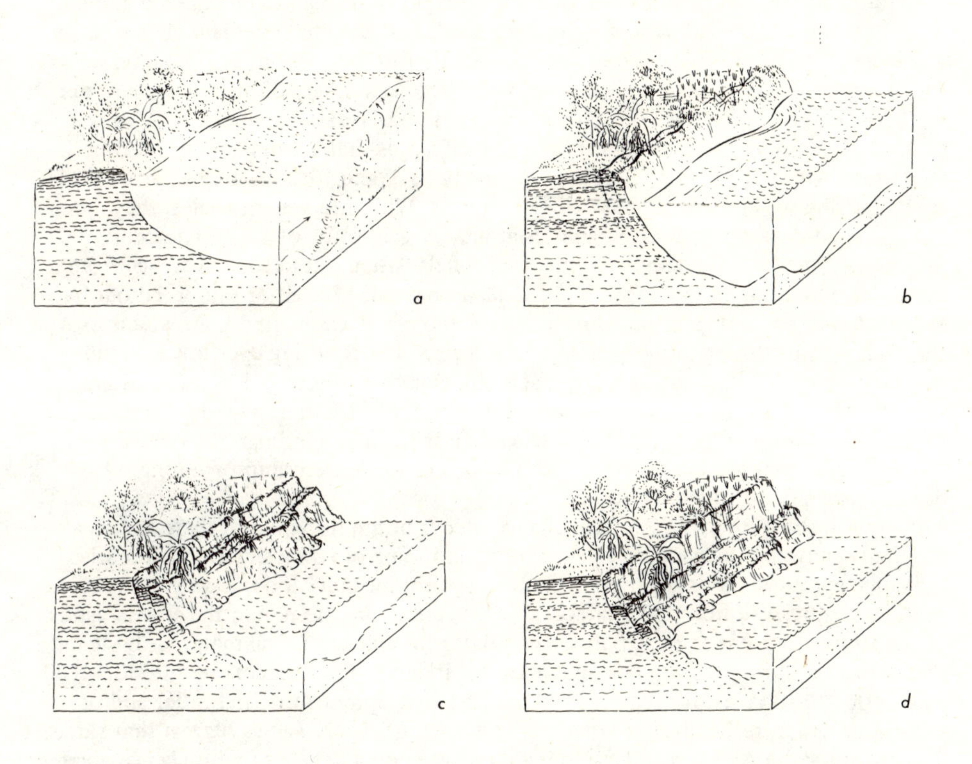

Fig. 10. Terras caídas. a and *b* - deepening of channel by scouring action of eddies is main cause of bank caving, which is announced by cracks running parallel to the water's edge; *c* and *d* - stability of the transverse profile reestablished by slumping in the direction of scour.

Deposition is accomplished in two major ways: lateral accretion occurs within the channel by bankside shoaling; vertical accretion, when floodwaters drop their load outside of, but mostly close to the channel. Overbank deposition builds up a strip of higher land, a natural levee (*restinga*), along the margins of the channels. Depending on the local amplitude of river stage, the crest of the restinga may lie several meters higher than the permanently-flooded backswamps or *igapós,* away from the river.

When a stream concurrently cuts into one bank and builds up the land on the opposite margin, the channel shifts across the floodplain, demolishing precisely the higher river-front tracts. If at a given place overbank deposition cannot hold its own, relative relief between levee and bottomlands ceases to exist as the river encroaches directly upon the low backswamps. But if the rear of the levee under attack is built up at flood stage by vertical accretion as its front crumbles into the river, if construction can keep pace with destruction, the levee moves ahead of the advancing river.

Although the processes just described occur in all floodplain rivers, the patterns delineated by the shifting channels may vary considerably (*Fig. 11*). Unlike some of its tributaries (*e. g.,* the Juruá), the Amazon does not have a meandering course. Its waters fork and come together again, embracing lenticular islands, which, strung along the river for many hundred kilometers, split the stream bed into a master channel and one or more side ducts, called *paranás* (*Fig. 12*). Vertically accreted overbank deposits, being thicker riverward, give a saucer-like cross-section to each island, in the center of which shallow lakes expand and contract with successive stages of high and low water.

Whereas in the meandering streams the pattern of quasi-concentric growth arcs suggests the coils of a half-wound spring, alluvial ridges that are built up and left behind as the Amazon and the paranás sway slowly back and forth in their courses, are gently curving or, in some reaches, almost straight. With intervening swales, they form a characteristic washboard topography that may be said to be of the "first order", since it is formed by the activity of major channels. When left at some distance from the active, migrating channel, this relief tends to become effaced or smoothed out by sedimentation and compaction. During floods, the rising waters find their way into the backswamps through low sections in the levees. The resulting overflow channels, occupied every year by muddy waters, actually constitute more or less intermittent streams, which assume a sinuous pattern and are flanked by their own deposits that form a topography of "second order" (*Fig. 13*). With their windings and ramifications, the strips of higher ground partition off portions of the backswamps, forming subbasins seasonally occupied by lakes.

In respect to the rate at which changes occur in the floodplains, there is a great difference between the gangliform Amazonas and some of its meandering tributaries. Erosional episodes attendant upon the cutting-off of meander loops have contributed to the widespread notion that "cataclysmic avulsions" are the rule in the alluvial lands of Amazônia. In point of fact, radiocarbon dates for archaeological materials lying on or close to the surface of the floodplain on one island in the Amazon are: $2{,}050 \pm 120$ and $1{,}100 \pm 130$ years, for materials found on topography of the first order and of the second order, respectively (Broecker, *et al.,* 1956). These values suggest that várzea lands flanking the Amazon and its lateral paranás enjoy a degree of permanence greater than that suggested by current ideas on the instability of alluvial formations in the region (Sternberg, 1960).

Annual Rise and Fall of the River. On the várzea, life--man, beast, plant--is at-

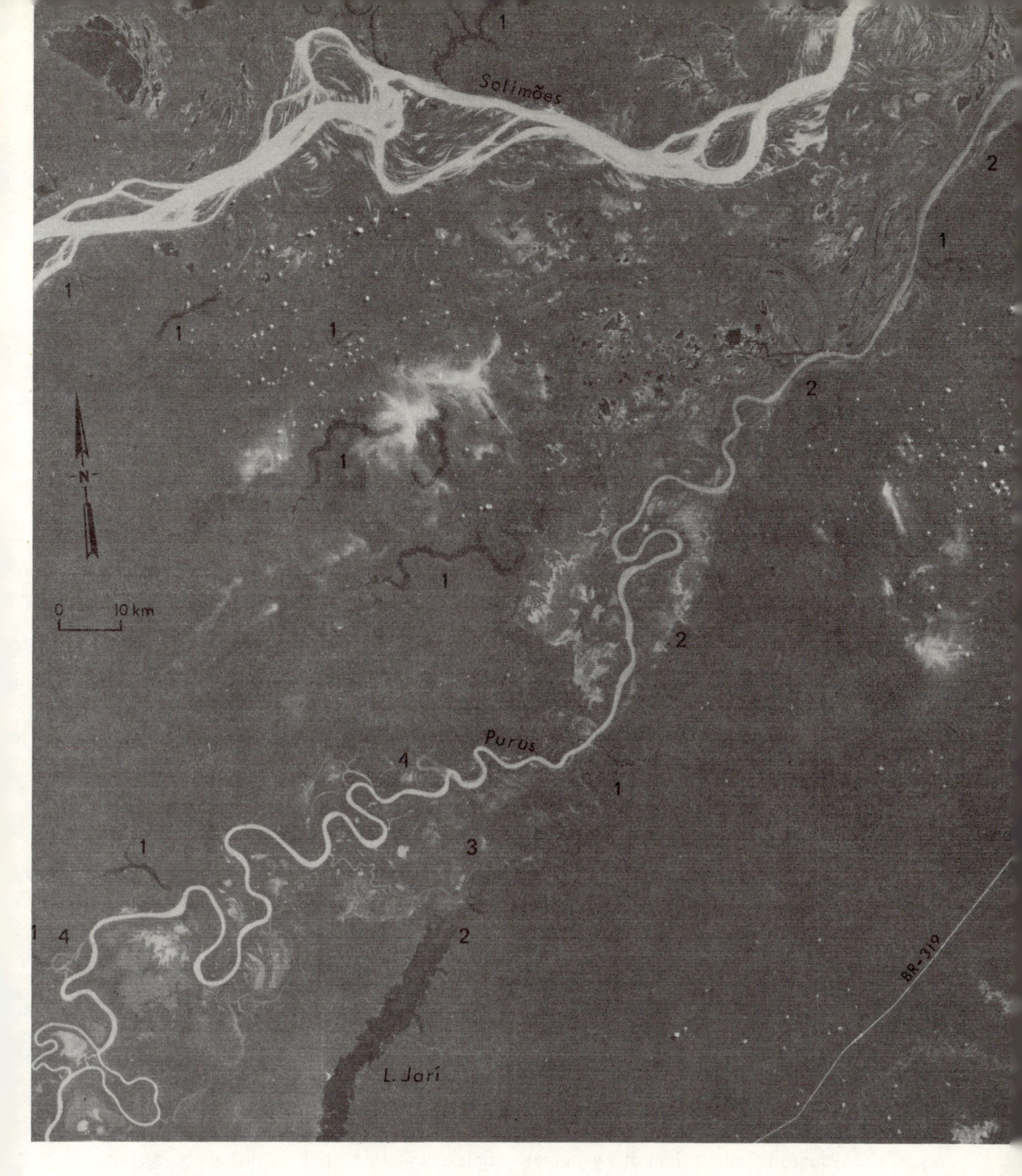

Fig. 11. Diversity of drainage patterns in the Amazon Plain. Features incised in terra firme, including L. Jari, are controlled by joints and fractures, (some other examples, numbered *1*). Upland bluffs (*2*) that constitute the east bank of L. Jari extend northeastward, bounding Purus floodplain and restraining eastward migration of this river's meander belt. Emissary (*3*) from L. Jari also receives water from Purus, into which subsequently it debouches; in fact, it meanders on fill deposited by Purus when a section of this river followed the alignment of the lake. Ganglionated and relatively stable pattern of Solimões is very different from that of rapidly changing Purus (examples of oxbow lakes: *4*). (NASA ERTS-1 Image, 31 July 1972).

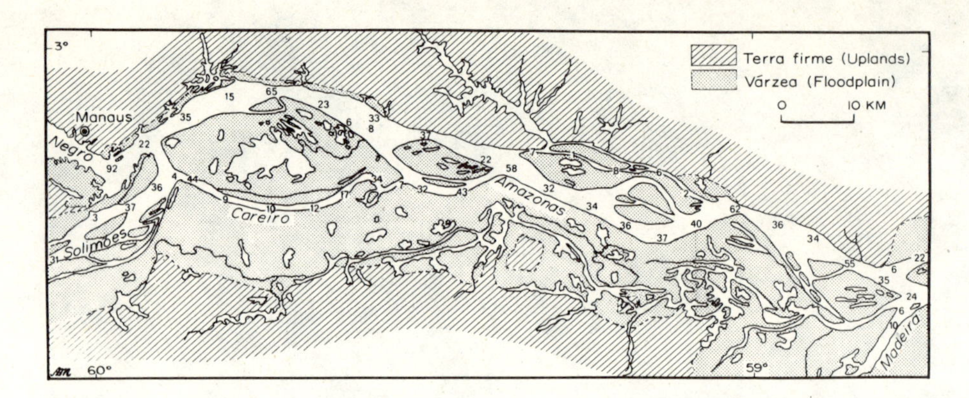

Fig. 12. Stretch of Amazon River, showing how lenticular islands divide the stream-bed into a master channel and one more side channels (*paranás*). Selected soundings (expressed in meters; datum: mean of exceptionally low stages) from the 1967 hydrographic survey by Brazilian Navy; taken, together with most water surface lines, from *Brazil – Rio Amazonas. Carta de Praticagem* Nos. p4 106 A and B, scale 1:100,000, 2nd. ed. 1970. Areas beyond scope of chart, including contour of terra firme lakes and approximate boundary between uplands and floodplains, from aerial photographs.

tuned to the pulsation of the waters that lay claim to the land during the annual high flow or fall back to their channels at low stage. It is a mighty heartbeat, and a steady one: the inundation that covers much of the floodplain, year in and year out, is remarkable, not only for its magnitude, but also for its regularity.

The Amazon runs roughly parallel to and not very far from the equinoctial line; left-bank tributaries rise in the northern, right-bank affluents in the southern hemisphere. Considering that summer rainfall maxima typical of the tropics prevail, six months apart, on both sides of the equator, one might *a priori* conceive perfect complementarity in tributary inflow. Given comparable volume of runoff, as well as reciprocity of high and low water stages between northern and southern affluents, the Amazon would, in fact, have minimal amplitude. Except at its mouth, however, the trunk stream flows south of the equator and only a fraction of its catchment lies north of the line. Even tributaries that reach into the northern hemisphere do not, in general, go beyond some 4° lat. and, therefore, tend to be more characteristic of an equatorial rather than of a genuinely tropical climatic belt. They do, of course, contribute to reduce the difference between peaks and troughs in the Amazon discharge--the high/low flow ratio of about 5 to 1 is significantly inferior to that of other large rivers (USGS, 1972). But the pattern of seasonal variation in the discharge of the lower Amazon is strongly influenced by its farflung southern tributaries, slowly traversed by late summer/early fall flood waves. Downstream from the mouth of the Tapajós, at Taperinha, high water occurs in May, low water in November (computed from unpublished data, courtesy Divisão de Águas, Departamento Nacional da Produção Mineral). At the confluence with the Negro, the waters rise gradually for some 8 months from a

Fig. 13. Alluvial features of first and second orders. Overbank deposition builds a narrow strip of higher land, *restinga,* along the margins of sediment-rich streams. Here, the Amazonas (upper right corner) has swept northeastward across the field of the photograph, leaving behind a series of arcuate, subparallel restingas, with intervening sloughs occupied by swamps or lakes. The ridge and swale landscape created by the Amazon and its paranás as they sway back and forth in their courses constitutes an alluvial topography of the "first order." Distributary channels, which carry silty waters into floodplain lakes that are part of the first order features, often assume a more or less perfect meandering pattern; they build their own flanking deposits and partition off sub-basins in the lacustrine millieu. The resulting features, well exemplified in the northeast section of the photograph, are of the "second order." Water about 1 m above mean-stage level. (Cruzeiro do Sul, S.A.).

minimum in October or November to a maximum in June, and drop in about 4 months to the succeeding low (*Fig. 14*). Average annual fluctuation at this point, as measured on the Manaus gage, some 18 km up the Rio Negro estuary, is 10 meters, and the difference between recorded maximum (1953) and minimum (1963) is 16 meters.

The already-mentioned regularity in annual water-level fluctuations becomes obvious when stage hydrographs plotted by decades are compared. The immensity of the total catchment area (of the order of 6,000,000 km^2), the very gentle gradient displayed by the Amazon (less than 100 m in more than 3,000 km) and by lowland sections of its tributaries, the vast capability for temporary storage of várzeas and affluent estuaries--all these are stabilizing influences that contribute to dampen short-period events.

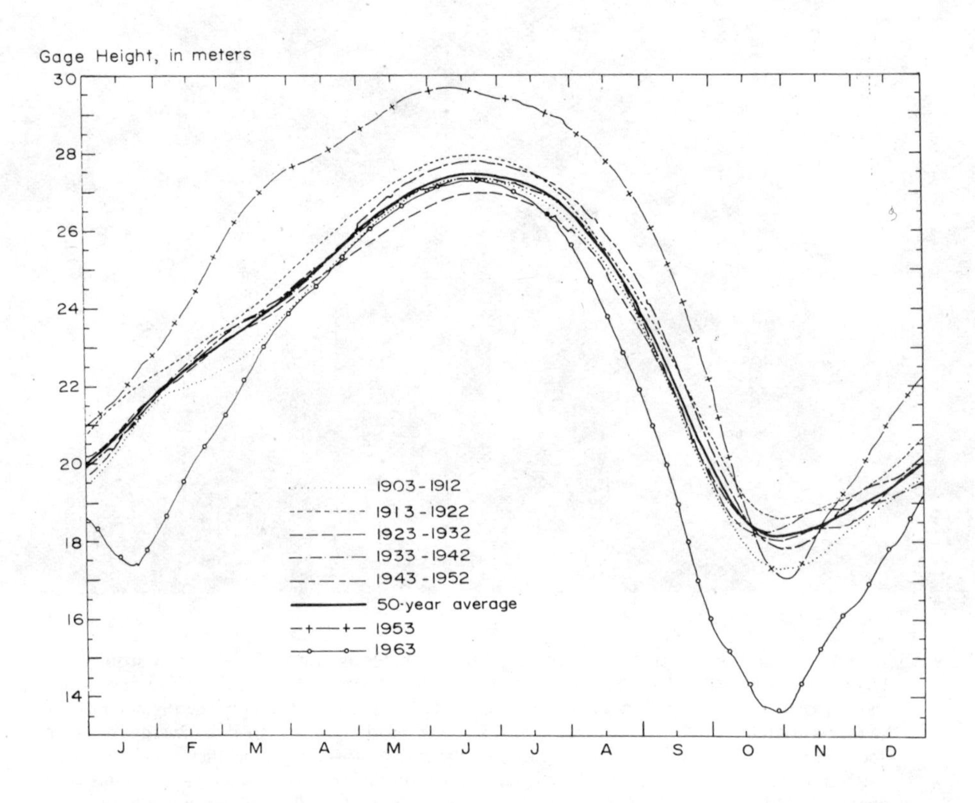

Fig. 6. Daily river stages at Manaus for 1903–1952, average and by decades; and for the years of maximum (1953) and minimum (1963) recorded stages. The great regularity of seasonal flows is striking. That the zero of the staff gage corresponds to mean sea level is only an assumption based on barometric observations made many years ago. Source of data: Manaus Harbour Ltd.

Ecological Significance of Variation in Water Level. The seasonal invasion of vast stretches of várzea by floodwaters is, in multiple ways, of great ecological consequence. It provides, for instance, an opportunity for seed or fruit to bob away, literally leaving behind pests and predators that tend to concentrate near the parent tree. The importance of water as the prime agent in the dispersal of várzea trees was emphasized by Huber in 1905. Many plants in the highly-diversified floodplain communities have evolved successful adaptations for this means of emigration, as pointed out by Ducke (1949), who compared species of the same genera belonging to bottomland and to upland habitats. Thus, tissue that provides flotation, or pods that remain closed at maturity (with terra firme counterparts dehiscent) are adaptive in areas subject to inundation.

Variation in river level is no less important for aquatic and semi-aquatic vegetation. In the case of white waters or of clear waters seasonally enriched by siltbearing floods, penetration of radiant energy is blocked by the very turbidity that denotes the presence of mineral nutrients. With photosynthetic activity thus limited to the topmost layers of water, primary productivity in the form of algae and of permanently submerged macrophytes is much reduced. Floating plants appear as optimally adapted: regardless of water-level oscillations, they remain in the region of active photosynthesis, where they use energy available at the surface, but not in depth. Small wonder that vast areas are covered by so-called "floating meadows" (Gessner, 1959). Although these are made up mainly of grasses (notably *Paspalum repens* and *Echinochloa polystachia*), the root mats (sometimes more than 1 m thick) may subsequently be colonized by woody, non-floating species, even robust trees. A discussion of some effects of changing water level on aquatic and semi-aquatic forms is found in Junk (1970).

The periodic appropriation of the várzea by floodwaters has profound implications for the aquatic fauna of Amazônia. Thus, for instance, during the seasonal spawning run, fishes pass freely into lakes and flooded bottomlands, which serve as breeding grounds. Furthermore, there is reason to believe that an appreciable part of the food intake of fishes and other components of the Amazonian river fauna (*e. g.* turtles), stems from terrestrial environments; studies of gut content reveal the presence of insects, seeds, fruits and the like. The "browsing" area of such organisms is expanded during the high water stage, when the forest is invaded by the floods. Indeed this is the time when many species develop the fat stores (Huber, 1910) that will carry them through the "physiological winter" of the low water season (Lowe-McConnell, 1967; Marlier, 1967).

It must be remembered that plants and animals evolve as members of ecosystems. Incidentally, when indigenous peoples of the Amazon, who live close to nature and are attuned to its manifestations, recognize as a taxon, say, those plants that produce "tambaqui fruits" (Verissimo, 1895), *i. e.* those savored by *Colossoma bidens,* they are translating their perception of the fact that a given fish and certain trees belong to a system. Huber seems to have been the first to ascribe a role to phytophagous fish in the dissemination of Amazonian plants. He referred to several large seeds found in the stomach of an unidentified fish, purchased in the Belem market, adding that they sprouted when planted. Ducke (1949) has suggested that the frequent tartness of floodplain fruit, in contrast with insipid or sweet counterparts of upland species, may have been selected for by the riverine fauna that eats them and passes the seed unharmed, thus contributing to dispersal. The correlation takes on an additional mutualistic dimension in the light of recent evidence that, in the course of evolution, at least some fishes have lost the capacity to synthesize ascorbic acid (Chatterjee, 1973). Consequently, however large or small their needs, they must get it from the environment--and the acidulous fruits of the várzea would seem a likely place to start tracking down the Amazonian fishes' dietary source of vitamin C.

In the light of the preceding observations, proposals for building dams across the Amazon (Panero, 1967a, 1967b) or major tributaries must be viewed with even greater concern than would otherwise be the case. The creation of stagnant bodies of water in an inner-tropical milieu is likely to set off a chain of harmful effects. The nature of these is suggested by some of the "ecological boomerangs" that resulted from the construction of large reservoirs in such an environment (Scudder *et al.,* 1972). By eliminating floods, dams in the Amazon might, in addition, wreak irreparable havoc on the icthyological fauna, whose evolutionary history seems quite different from that of its African counterpart and has resulted in few typically lacustrine species (Marlier, 1973).

Cultural Significance of Variation in Water Level. When one considers the rise and fall of the river in terms of human occupance, certain conspicuous features come to the mind of even the most casual observer. For instance, how the people of Manaus cope with variation in water level: the floating stores; the houses on tall piles that line the urbanized rias; or the port's two floating piers (*Fig. 15*).

However it is on the várzeas that the web of relationships between man and the most dynamic component of the environment attains greatest complexity. Harvesting of native flora and fauna, farming, and transport of farm products to market--all move to the beat of the river.

Waters that create and constantly retouch the várzea farm tracts also affect the manner in which the settlers distribute the use of their holdings, both in space and in time. To the degree in which floods saturate, submerge or leave the land dry, they not only condition the extent of farmland that may be brought into play, but also the time span during which the lands may be put to use and even the manner in which such utilization is possible (*Fig. 16*). Bananas, rubber and cacao, for instance, are generally

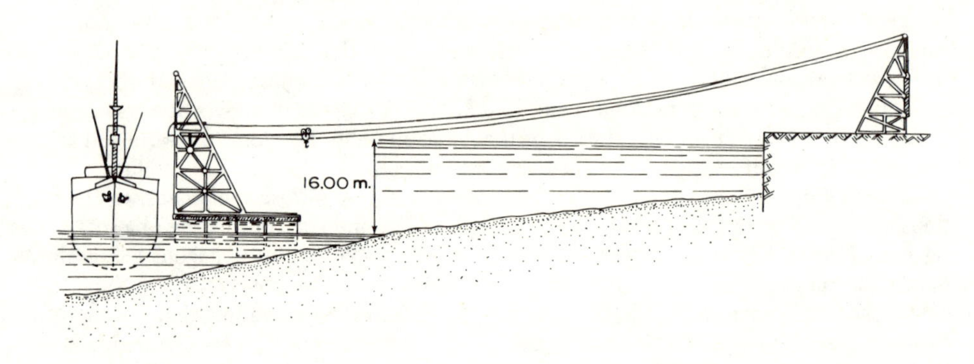

Fig. 15. Port of Manaus. Section at floating pier, built during the rubber boom and servicing the ships berthed alongside by means of cableways attached to towers on land. A second floating pier, not shown, is connected to shore by a floating trestle ("roadway")

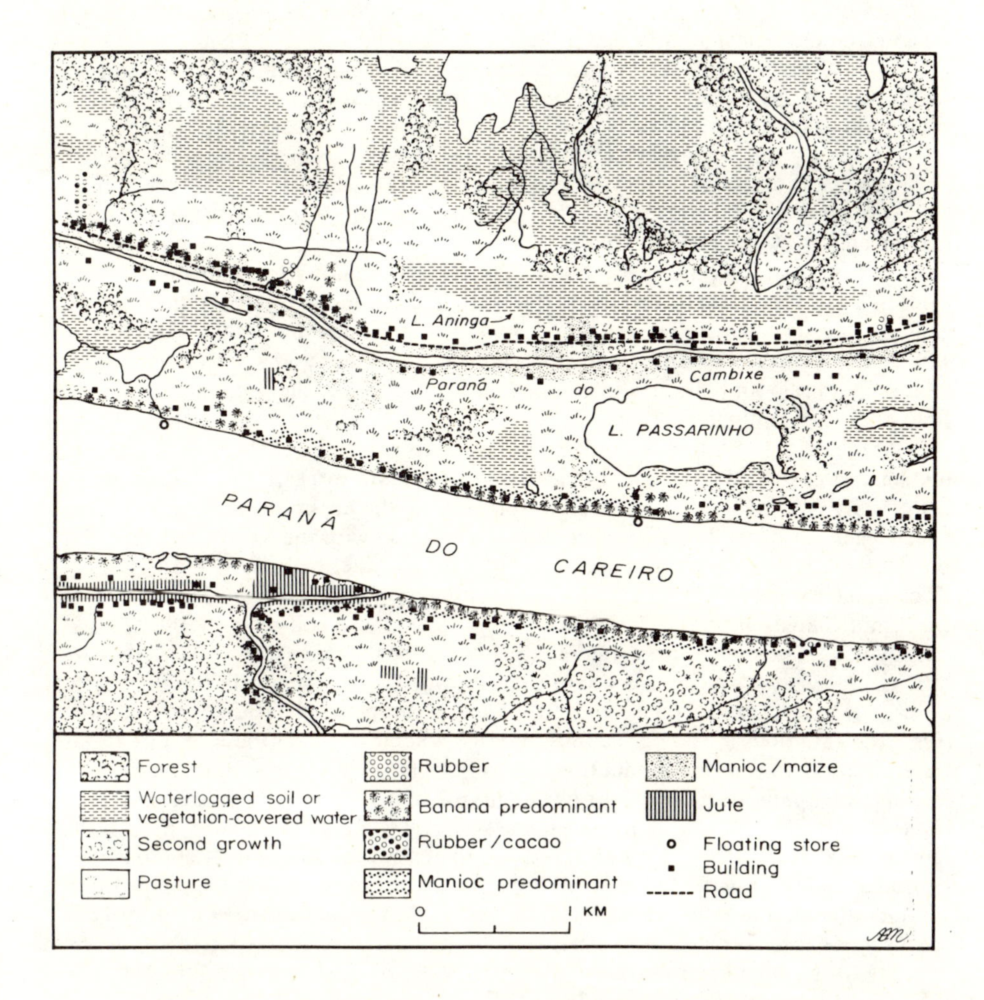

Fig. 16. Land use along the Careiro and Cambixe channels, near the confluence of the Rio Solimões and the Rio Negro. The distribution of various crop combinations (in 1956) is indicated in a generalized manner by segments of streambank, not by individual establishments. The representation of some crops, such as beans, sweet potato, rice and vegetables, which are significant on certain farms, is not compatible with the degree of generalization of the map. Fruit trees are disseminated throughout most properties and some stretches of the paranás are lined with mango trees.

found on the better soils of the restingas. Jute often occupies the somewhat heavier soils to the rear of the natural levee; not only are higher areas reserved for other crops, but slack water, such as is found in the bottoms, is required for retting the stems. Maize occupies an intermediate position: here, alongside jute; there, contiguous to manioc or cassava (the regional staple), which prefers higher ground.

That the agricultural calendar is geared to the river can best be seen at rising stage. As the strip of unsubmerged land shrinks, the growing season is cut short and the farmer must work against time to harvest and, as in the case of manioc, process his crops.

In some parts of the várzea, a distinctive cattle industry makes periodic use of low-lying tracts that otherwise would be turned to little account; by felling and burning backswamp forests, grazing lands supplementary to those held on the river front may be created. Cattlemen also make good use of the "floating meadows" as the main source of forage during the yearly floods. Inundations do not affect equally all grazing lands in a given section of floodplain. A few centimeters more--or less--in the height of alluvial deposits make all the difference. Some farmers know that even a modest flood will deprive them of dry land upon which to place their cows; others who dispose of ground that has been built up slightly more suffer hardly any damages, unless the overflow reaches exceptional levels.

Herds raised or fattened on the várzeas--the total may be of the order of 1/2 million head--consist mostly of beef cattle. Some areas, however, carry a considerable number of dairy cows. The prime example is found along the banks of the Paraná do Careiro, which leads out of the Solimões, just above its junction with the Negro, and flows into the Amazonas some 40 km downstream (*Fig. 12*). Together with an ancillary channel of its own, the Paraná do Cambixe, the Careiro is the chief component of the Manaus milkshed (more details in Sternberg, 1956, 1966). Production is collected in 50-liter milk cans by daily mixed passenger-cargo boats. Because of the stress caused by flood conditions, especially when the cows have to be crowded on pile-supported or floating platforms, *marombas* (*Fig. 17*), milk output is likely to drop sharply at times of high water (*Fig. 18*). Variations in stage affect not only the volume of milk produced, but also its transportation. Part of the production comes from farms along the Cambixe; if the level of water permits, it is brought out by motor launch. In the dry season, however, not even a canoe can navigate the entire passage, and great effort must be made to ship out the milk. Some of the dry-season output, therefore, remains in the Cambixe and is converted to butter or cheese, in small domestic plants. Nevertheless, overall prospects for milk production on the várzea seem favorable. Plans have been drawn up for two specially designed refrigerator boats, complete with pasteurization equipment on board, to make six- to eight-day runs out of Manaus (1970 pop. 284,000) and thus expand considerably the area of the present milkshed (Ecotécnica, *ca.* 1967).

Fig. 17. A small improvised *maromba* on piles contains 25 head of *várzea* cattle during the greatest inundation on record (1953). Canarana (*Echinochloa polystachia*), the preferred flood-time forage, fills one of the boats; it is often gathered in distant "floating-meadows." Water level 6.31 m above mean stage. (H. O'R. S.).

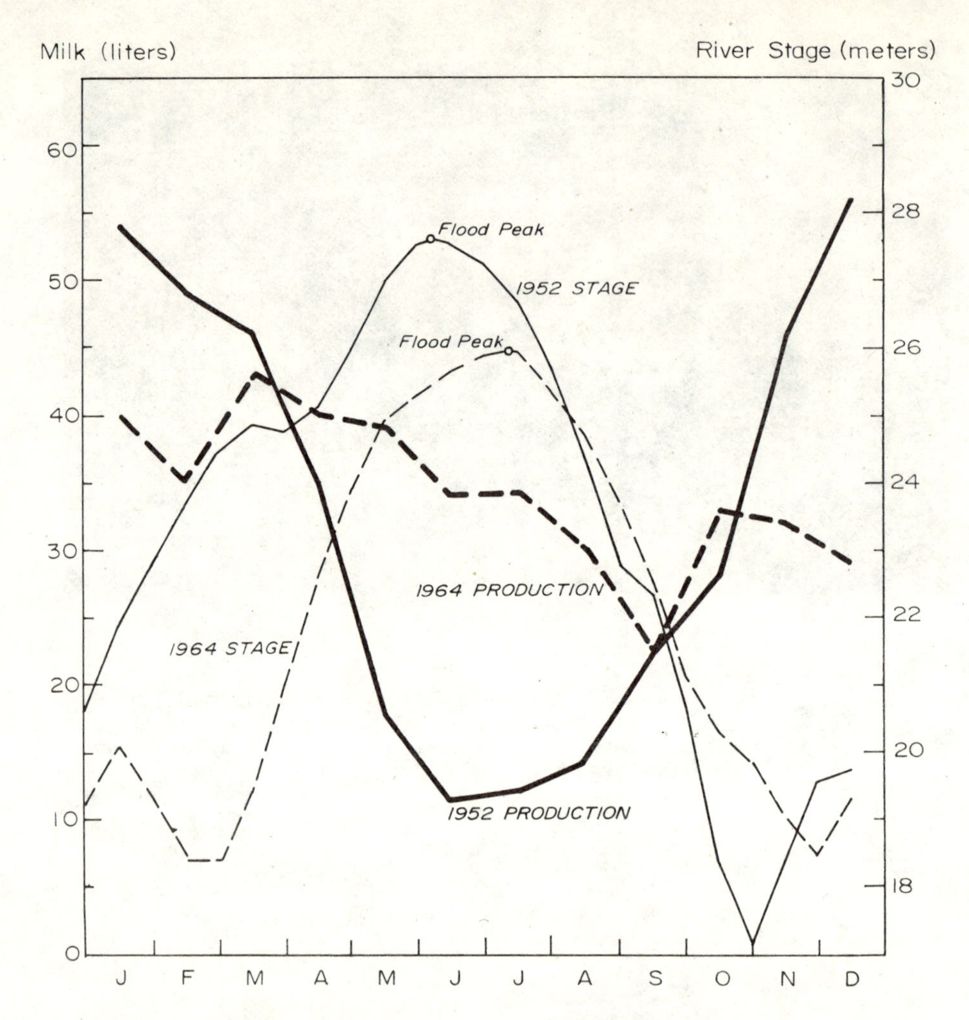

Fig. 18. Average monthly production of milk per cow on a floodplain farm, Fazenda "Carnaubinha", Paraná do Cambixe (see Fig. 16), and average monthly river stage, for 1952 and 1964. The 1952 flood peak coincided almost exactly with the long-term average high water stage; as the waters receded that year, milk production made a good comeback. In 1964, with a lower than average flood peak, there was not the same rapid recovery. Dairymen, while dreading a "great" flood, speak of the value of a "good" flood, which fertilizes the pastures and, supposedly, helps to keep some pests under control.

TRIBUTARIES AND LUSO–BRAZILIAN EXPANSION

Despite controversies regarding the precise position of the Meridian of Tordesillas, agreed upon in 1494 to delimit the possessions of the two Iberian powers, there was little doubt that, at best, only a narrow wedge at the mouth of the Amazon could rightfully pertain to Portugal.

However Dutch and English "heretics" established plantations and factories in the lower Amazon region at a time when the Iberian peninsula was unified under a common sovereign (1580–1640). The Council of State of Spain, not without some trepidation, entrusted to the Portuguese the task of dislodging the interlopers and securing the Iberian hold on the valley. Spanish orders were clear: regardless of who actually did the fighting, "this conquest is by the Crown of Castille" (Reis, 1947).

The stipulation was of no avail. Having routed English and Dutch with the aid of native allies, the Portuguese pushed up the Amazon and major tributaries. Explorers, soldiers, traders, slavers, missionaries (of the Franciscan, Carmelite, and Jesuit orders, among others), fanned out along the dendriform river system, effecting an inexorable headward erosion of Spanish *de jure* territory. By mid-18th century, a ring of fortifications, strategically planted on remote reaches of the affluents, staked out the *de facto* possession of Portugal, a gigantic amphiteatre that foreshadowed the present contours of Brazil's share of the Amazon basin.

Whether a given affluent became a corridor of circulation and settlement or remained for centuries a more or less unchanged sanctuary was determined perhaps less by such natural obstacles as rapids and falls, as by the riches thought to exist along or at the end of the watercourse--and by geopolitical considerations.

It is in the context of the preceding paragraphs that one should view the events that took place on the Xingu, the first major right-bank tributary encountered upon ascending the Amazon. At the close of the 16th century, merchants from Flushing built two fortified trading posts, Orange and Nassau, on the banks of its lake-like estuary (*Fig. 19*). This early attempt by Europeans to organize permanent trade with the Indians of the interior was cut short in 1623, when the Portuguese laid waste the setlements (Edmundson, 1903; Williamson, 1923).

In the meantime, diverse indigenous groups, showing little desire to deal with the white man, were retreating into the headstreams of the Xingu, on the oldlands of Central Brazil. Here was an area segregated by an extensive stretch of falls and rapids and surrounded by rather poor savannas, which until recently aroused little interest. Never made famous through a rich strike in minerals or a boom in some forest product, the Upper Xingu attracted few trespassers until the end of the 19th century (Galvão and Simões, 1964).

The headwaters, which drain a bevelled, sediment-covered surface, converge on the main stem of the Xingu in a pattern that has been likened to a "feather duster". (*Fig. 20*).This provided the Indians with what must be at least 1,500 km of stream channels; during the rains, overflow increased the range and mobility of the jatobá-bark canoes. The ingathered tribes, representing major South American linguistic stocks (*e. g.*, Tupi, Gê, Arawak and Carib), having interacted for centuries over these waterways, produced, despite language differences, a unique amalgamation, a classic example of intertribal acculturation: the so-called "Xinguan" culture (Schaden, 1969). The aboriginal population in the region has dwindled greatly, there being probably less than 1,000 (Villas Boas, 1968, 1973) who live in the Xingu National Park, an 80 km-wide belt

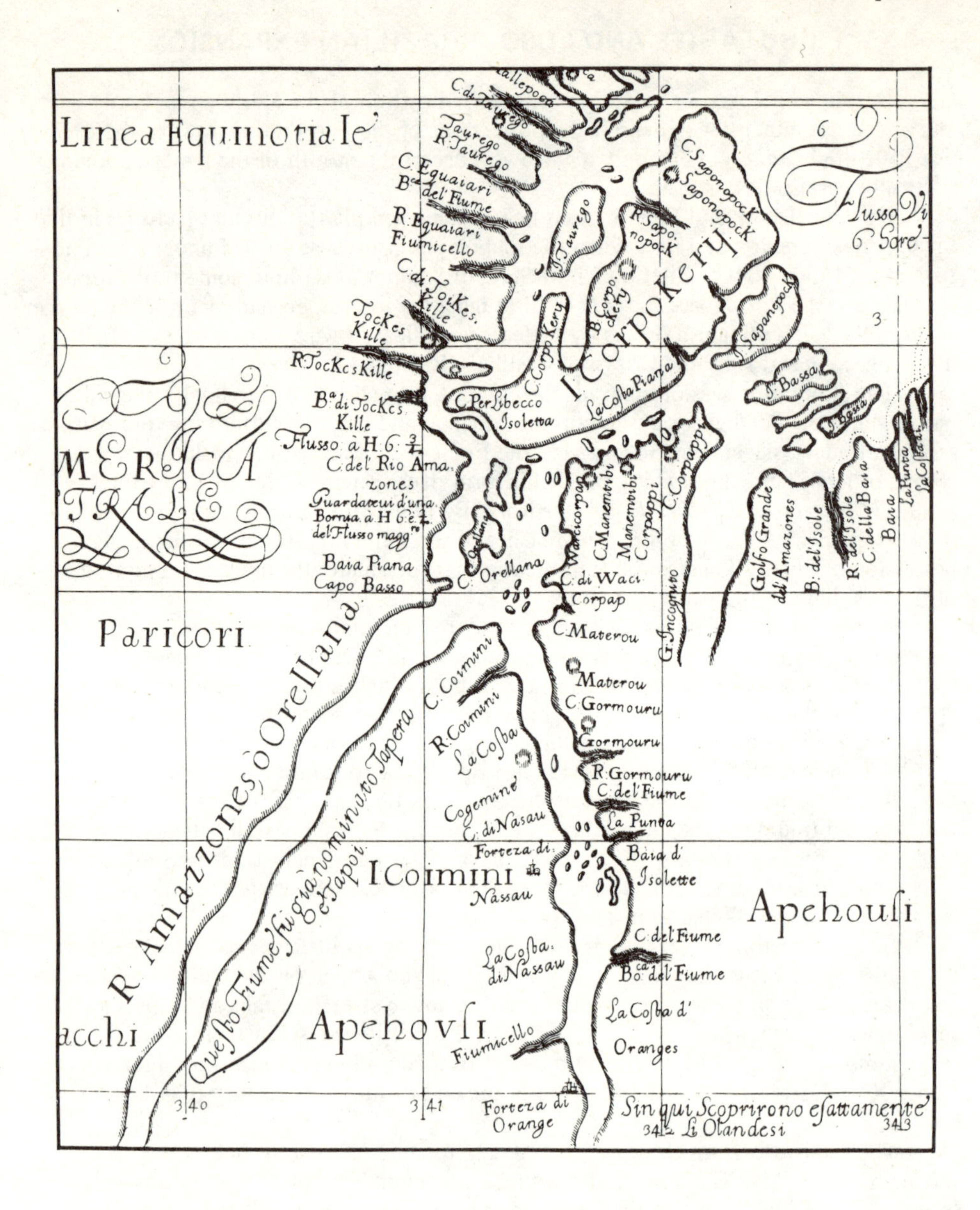

Fig. 19. Early exploratory and trading activities of the Dutch, inland from the mouth of the Amazon, are reflected in the annotation along the bottom edge of the map, indicating "exactly how far the Hollanders discovered" the Xingu river. Note also the designation of the east and west bank of this stream ("The Coast of Oranges" and "The Coast of Nassau", respectively) and that of the fortified trading posts on the left bank. Extracted from "Carta particolare dell'rio d'Amazone con la costa sin al'fiume Maranhan", published in Sir Robert Dudley's *Dell'arcano del mare* (1st ed. 1646–47). The first two of three manuscript volumes, formerly preserved in Florence and which may have served as a basis for the printed work, are dated 1610.

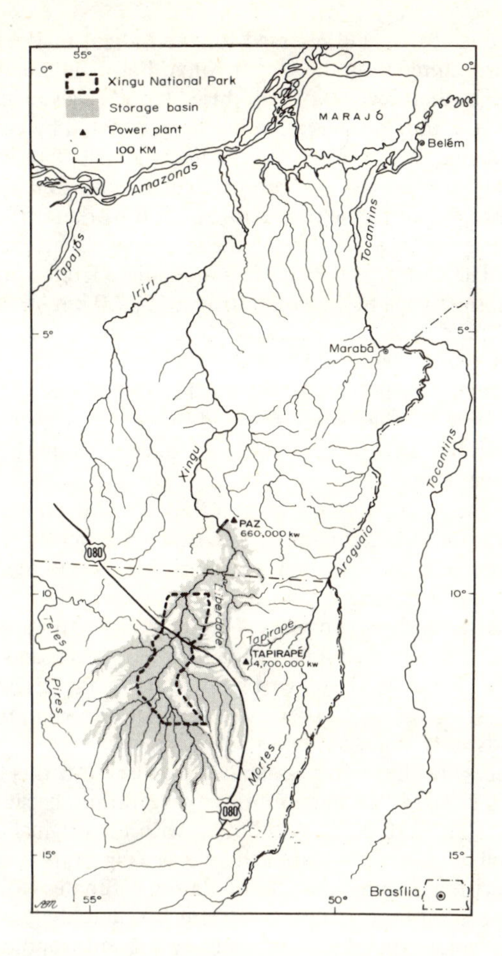

Fig. 20. Proposed Lake Paz. Based on the premise that, at the same latitude, the bed of the Araguaia River lies some 100 m below that of the upper Xingu, a proposal has been advanced to build a dam on this river at Paz Island. The water of what has been promoted as the largest artificial lake in the world, would back up in the valley of the Liberdade and, by way of the Tapirapé, spill into the Araguaia. Plans call for one hydroplant at the dam, another at the diversion. The 60,000 km^2 impoundment would drown most of the Xingu National Park, established in 1961 as an Indian sanctuary, but already violated by highway BR–080 (of which only the pertinent section is shown here).

placed astride the trunk stream and reserved for the Indians in 1961. Immense cattle ranches are now encroaching upon the Upper Xingu Basin, an area that once may have appeared to the Tupi as their legendary "promised land" (Galvão and Simões, 1964). Even the integrity of the Indian reservation has been violated by current road construction programs (Junqueira, 1973). When federal highway BR-080 ripped through the Xingu Park, it not only drove away a group of Txukahamãe; fourteen of these Indians died of measles contracted from the road-construction workers. The charge also has been made that the road has exposed the Park to the action of squatters and landgrabbers (Villas Boas, 1974). The danger is perhaps less immediate from a proposal for hydroelectric facilities whose storage basin would drown 60,000 km^2 (Almeida, 1967; Burnier, 1971), an area more than double that of the reservation (*Fig. 20*)--most of which would be covered by the impounded water. Although the project appears to have been tabled for the time being, it will hang as yet another threat over the Indians of the Xingu. If and when more detailed studies should prove the project to be technically feasible and economically attractive, the mere presence of native Brazilians is not likely to deter its implementation.

Proceeding up the Amazon from the mouth of the Xingu, the next major tributary encountered, also on the right bank, is the Tapajós. When a group of Dutch or English colonists sailed into the drowned mouth of this river in the early 17th century, with the idea of establishing tobacco plantations, they were forced to flee after many of the party were massacred by the natives. These belonged to the historic Tapajó nation, after which the river is named--one of the most numerous and organized sociopolitical bodies in Amazônia. Heriarte, writing in 1662 of the "*Provincia dos Tapajos*", referred to the "largest village and settlement . . . so far known in this district" and claimed that "it fields sixty thousand bows when it makes war." Nimuendajú (1949), who did not have access to the Heriarte manuscript, discounts this figure: "a misprint" of the text published in 1874 "or an enormous exaggeration, because it would presuppose a population of some 240,000." A collation with the original (Heriarte, 1964) shows there was no typographical error. Whether there was exaggeration is not known, but if so, recent upward revisions of pre-contact population estimates (for instance, Dobyns, 1966; Borah, 1970) suggest that it may have been considerably less "enormous" than appeared to Nimuendajú, writing in 1939. Actually this anthropologist does not deny that "the vestiges of ancient settlements point to an exceptionally numerous population," and himself discovered the site of 65 villages in the vicinity of Santarém, at the mouth of the Tapajós. With the recognition a century ago that "*terra preta de índio*" (Indian black earth) marks the premises of ancient settlements along the Tapajós (*e. g.* , Hartt, 1874, 1885), came the realization that the bluffs must have been lined with these villages, for the black land is almost continuous" (Smith, 1879). And, furthermore, that villages "must have stood upon these spots for ages, to have accumulated such a depth of soil about them" (Brown and Lidstone, 1878). Potsherds are found throughout the terra preta (which may be as much as 1.5 m thick); in places they are so plentiful it is claimed they interfere with cultivation (Hartt, 1885; Barata, 1950). Much of the pottery is of the elaborate Santarém or Tapajós style, of which one of the most highly developed forms is the caryatid vessel (see the excellent illustrations in Palmatary, 1960).

The *terras pretas,* one may note, are not limited to the Tapajós, and are found on floodplains (Sternberg, 1960), as well as uplands, where, "highly prized as agricultural grounds" (Brown and Lidstone, 1878), they present a marked contrast to the prevailing poor, yellow latosols. It is remarkable that in an environment such as Amazônia, whose potentials have been judged insufficent to support large concentrations of population or stable settlement (Meggers, 1954), indigenous settlements should have been so large and persistent--leaving aside the matter of cultural achievement--that soil scientists have come to recognize the anthropogenic black earth as a "taxonomic unit" (Falesi, 1967).

In 1639, a Portuguese force broke the back of the Tapajó; by the end of the century they seem to have disappeared as a tribal unit (Nimuendajú, 1949) and were considered possibly extinct in 1820 (Spix and Martius, 1823–31), "*spurlos verschwunden*" (Martius, 1867). Such massive depopulation, as elsewhere in the Amazon, resulted from slave-hunting forays and the ravages of Old World diseases in an immunologically virgin people. Ironically, when in 1929 the Ford Motor Company established a rubber plantation on 1,000,000 ha along the once densely settled lower Tapajós, labor shortage proved to be a crippling handicap.

A transfusion into Amazônia of "surplus" population from Brazil's northeast was precisely one of the main objectives of the Transamazon. This pioneer highway, slashing inland from the eastern seabord, reached the banks of the Tapajós at Itaituba in 1972, those of the Madeira at Humaitá in 1974. Santarém (1970 pop. 51,123), to be developed as one of the key river ports serving this highway, will be directly connected to Itaituba, head of navigation, some 280 km upriver. Despite the rapids and falls that begin here, the Tapajós and its headstreams, the Arinos and Juruena, were long regarded by the Portuguese as potentially important in linking the Amazon with Mato Grosso, Brazil's Far West, and the La Plata drainage. Administrators, however, were frustrated in their efforts to encourage the use of the Arinos route, not only because of the falls, but also because travellers were in constant dread of the "multitude of Indians" (Oeynhausen, 1811), who lived along the upper, more isolated sections of the waterway. Even in the 1960's, launches plying the Arinos occasionally were showered with arrows by the remaining natives, who continued to resist the invasion of their territory.

THE MADEIRA PASSAGEWAY

The Madeira River was so named because of the quantity of timbers swept down every year by its waters. Towed ashore during a few weeks every season, great cedar logs (genus *Cedrela*) were enough at one time to support the year round operation of a sawmill at Itacoatiara, on the Amazon (Keller, 1875). Since the driftwood is derived from caving banks, its abundance serves to underscore the difference between the turbid, silt-bearing Madeira, on the one hand, and the clear-water Xingu and Tapajós, on the other. Whereas the latter have not been able to alluviate their estuaries, the former has choked its lower valley with sediment and, at its junction with the Amazon, contributes to form the more than 300-km long, low-lying island of Tupinambarana. The Madeira's headstreams do not merely evacuate the meager waste from Brazil's West Central Plateau; spreading out in a broad arc across the Cis-Andean Plains they also gnaw into the varied rocks of the youthful Cordillera.

Like the Tapajós and the Xingu, the Madeira provided a route between Pará and Mato Grosso, and offered the additional prospect of a rear entrance to the fabled Potosí region. In effect, on the west, by way of its affluents the Mamoré, the Beni and the Madre de Dios, it gives access to the "regions of silver and of gold, of chinchona and of coca" (Velarde, 1886). On the east, by way of the Guaporé, to the Mato Grosso gold fields (discovered in the early 18th century). Not surprisingly, the Madeira River became the object of considerable geopolitical maneuvering on the part of Portugal and Spain.

In 1748, the Portuguese detached the new Captaincy-General of Mato Grosso and Cuiabá from that of São Paulo. The Captain-General was instructed to establish the seat of government and take up residence in the Mato Grosso. This colony was to be made "so powerful that it impose respect upon and contain its [Spanish] neighbors and serve as a bulwark for the entire backlands of Brazil." The Overseas Council believed such an objective to be greatly favored by "ease of communication by water . . . with the city of Pará [Belém]."

It would be difficult to imagine a more remote and isolated outpost than Vila Bela da Santíssima Trindade, the new capital established on the banks of the Guaporé. The cross-country route from São Paulo required "six months of most troublesome navigation to Cuiabá and an additional month from there to Mato Grosso" (Corrêa Filho, 1969). As to the "providential" river connection to Pará, Vila Bela was, in fact, about 2,800 km by river above the mouth of the Madeira, more than 4,000 km from Belém--and some 8,500 km by combined fluvial and coastal navigation from the seat of viceregal power at Rio de Janeiro. Yet at least one 18th-century Governor, making light of the endless equatorial forest and the length and fragility of his life lines, strove for a style of graceful living that would hardly have been out of place in the Portuguese court itself. The walls of the Governor's Palace, frescoed "in the style of Watteau" (Fonseca and Almeida, 1899), looked down upon a succession of banquets, balls, masquerades, theatricals, poetry-readings and other entertainments, attended by "His Excellency and Nobility, attired in the richest silks" (Correa, 1777). The minuscule enclave of European civilization--and Portuguese power --enjoyed the rare privilege of a gold mint, and was estimated to number about 6,000 inhabitants at the end of the 18th century (Serra, 1858). Virtually deserted after the capital of Mato Grosso was transferred to Cuiabá in 1820, Vila Bela (today called Mato Grosso) shows a slow upturn in number of inhabitants during the 20th century: 340 in 1906; 828 in 1970 (*Fig. 21*).

THE RUSH TO THE "ALTOS RIOS"

As the Mato Grosso placers were worked out, traffic on the Guaporé and Madeira decreased. The Treaty of Madrid (1750) validated Portuguese possession of the east bank of the Guaporé and of both margins of virtually the entire Madeira. For a century, life in the riverine lands was unhurried, based on modest agricultural and gathering activities. In the second half of the 19th century, long after the former colonies of Portugal and Spain had gained their independence, the Madeira and its neighboring streams to the west became the scene of a feverish stampede that threatened to engulf the region in an all-out war. The lower and middle course of the rivers that rise in or breach the eastern flanks of the Andes are rich in rubber trees. Increasing world demand for gum elastic culminated in the great boom that drew into this area a stream of adventurers from Bolivia, Peru and Brazil.

The Brazilians, many of them refugees from disastrous droughts in the Northeast, followed the Amazon and its tributaries: a numberless counter-current of people flowing toward the *altos rios,* or "upper rivers" of western Amazônia, where productivity of rubber trees is greatest. They opened up and laid claim to what was Bolivian territory, even if mapped as "tierras no descubiertas" (Reis, 1937).

The most important thrust of the rubber tappers occurred along the Purus, where

Fig. 21. Vila Bela da Santíssima Trindade (later Cidade da Santíssima Trindade de Mato Grosso and now simply Mato Grosso). Founded in 1752 as the capital of the new Captaincy-General of Mato Grosso, Vila Bela was the terminus of the river route to Mato Grosso by way of the Amazonas, the Madeira and the Guaporé, on whose eastern bank it was built. The town, surrounded by swampy terrain, has a reputation of extreme insalubrity. After the seat of government was moved in 1820 to Cuiabá, a population drop from 5-6 thousand to a few hundred left Vila Bela almost a ghost town, but an upswing has occurred in the number of inhabitants and is accelerating as the result of a new highway linkage. (H. O'R. S.).

the first *seringal,* or rubber estate, was established in 1852 and steamship navigation was inaugurated in 1869. At this time, an immigrant from Maranhão, one Antonio Labre, founded a settlement on the Purus; in 1892 the rubber town of Lábrea (1970 pop. 3,017), is said to have had the greatest revenue among all intendances of Amazonas--including Manaus (Cunha, 1906). The primacy of the Purus River in output of rubber was assured by the quality and quantity of the product from its headstreams. One of these was the Aquiri or Acre River. By 1899, the Acre watershed supplied 60 per cent of the rubber produced in the Amazon drainage. It was this area, penetrated by tens of thousands of Brazilians, that became the bone of contention between Brazil and Bolivia. The conflict was brought to an end by a treaty signed in 1903, according to which Bolivia conceded the disputed territory in the headwaters of the Purus and neighboring Juruá. Annexed by Brazil, it later gave rise to the State of Acre (152,589 km^2), named after the most famous "rubber river" and comprising other contiguous headstream areas. As part of the settlement, Brazil pledged to construct a railway around the nineteen falls and rapids of the Madeira and its affluent the Mamoré, thus giving the landlocked Bolivians an outlet to the Atlantic. The 366-km long, 1-meter gauge Madeira-Mamoré railroad, which, at a high cost in human lives, linked Porto Velho and Guajará Mirim in 1912, has since become obsolete and is being replaced by a highway.

The Japurá (Caquetá, in Colombia) and the Içá (Putumayo, beyond the Brazilian border), north-bank tributaries that flow through the westernmost section of Brazil's Amazônia, like their southern counterparts, rise in the Cordilleran realm and carry a considerable load of sediment from the Andes and respective forelands. Their lower courses wind through alluvial plains and are free of rapids. The analogy between north and south tributaries disemboguing in the upper half of the Brazilian Amazon is heightened when one observes past resource use. Thus, it was also the rubber boom that opened up the Putumayo River, albeit with a savagery (Casement, 1911–12) for which one is unprepared --even when aware of the appalling exploitation in the *seringais* along the southern affluents. In the Putumayo, the indigenous population was enslaved to work as tappers, and eighty per cent were wiped out during the first decade of this century (Varese, 1972).

In contrast to the affinities existing between the north-bank and the south-bank tributaries of the Solimões that drain the Andean flanks, there is no suggestion of a mirror-image in the case of the left and right bank affluents that discharge into the Amazonas east of the Rio Negro; the clear- or black-water tributaries that issue from the Guiana shield are much shorter than their opposites, which head in the Brazilian plateau.

THE RIO NEGRO SYSTEM

The mighty flow of the Rio Negro is fed by runoff both from the forelands of the distant Cordillera Oriental and from the Guianan massif. The first discharge measurements ever effected indicated an outpouring of some 67,000 m^3/sec on July 22, 1963. This represented a contribution of some 40 percent to the aggregate volume of the Amazonas just below the confluence (Oltman *et al.*, 1964). The magnitude of the Negro's discharge is understandable in view of the fact that a large part of its course lies practically on the equator; precipitation figures for São Gabriel da Cachoeira, at 0°08'S, probably are typical, with about 3,000 mm per annum and, on average, 14 days of rain during the *driest* month.

Despite turbid inflow upriver, the lower course of the Negro is characterized by water that carries a minimal load of sediments and whose optical properties have astonished observers since the day when Orellana's party described it as being "black as ink."

The blackness of deep water that even at low stage plumbs more than 90 m in the wide estuary of the Negro is explained by the presence of a yellow-brown, water-soluble pigment, derived from decomposed plant material and enough to extinguish 8/10 of incoming light at a depth of only 20 to 25 cm (Kanwisher, 1967). As they flow together (*Fig. 22*), the darker, hence warmer, and sediment-free waters of the Negro, which hugs the left bank of the common channel, tend to be underrun by those of the Amazon. Turbulence at the sloping shear zone accounts for the burbles of white water that rise through and bloom on the surface of the overlying wedge of dark waters (*Fig. 23*).

Fig. 22. Confluence of Negro and Solimões rivers. Sediment-free and warmer "black" waters of the Negro, hugging the uplands to the north, are underrun by those of the faster, silt-laden, "white" waters of the Solimões (mean velocity in respective sections above confluence, during July 1963 gaging operations: 0.80 m/sec. and 1.20 m/sec.). Turbulence produced by shear at the interface, which dips towards the left bank, releases burbles of white water – isolated or forming continuous "curtains" – into the overlying wedge of black waters; the axes of these rising masses appear to lean in the direction of the left bank. Note at top center a series of bands of white water emerging suddenly in the midst of black water. Overbank deposition by Solimões has built up the densely settled rim of a lenticular island, partially shown on right. (Cruzeiro do Sul S. A.).

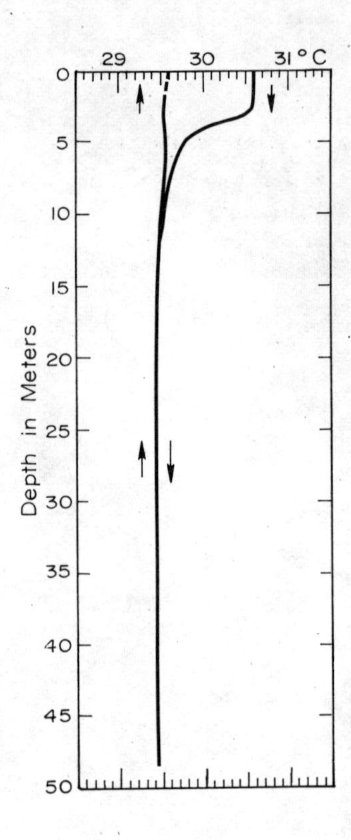

Fig. 23. Contrast in temperature between "black" and "white" waters at the junction of the Negro and the Solimões. A bathythermograph, lowered through a thin wedge of Rio Negro water, initially recorded a temperature of 30.6° C; at a depth of some 3 to 4 m it penetrated the underlying Amazon water, in which it continued during the remainder of its descent, registering 29.4° C. This temperature was more or less uniformly maintained as the instrument was drawn up through a rising boil of white water. From the original bathythermograph slide (November 26, 1963).

Optical differences are only the most conspicuous expression of profound dissimilarities in the character of the Negro on the one hand and of the Amazon (and its white-water tributaries) on the other. Indeed, prior to the actual discharge measurements made in 1963, differences in the amount of dissolved salts observed in the Negro, the Solimões and the Amazonas during late April and early May 1959 had been used to estimate the relative contributions of the two confluents. On this basis, the Negro was reported to be supplying about 40 percent, the Solimões about 60 percent of the Amazon's flow immediately downstream from the junction (Gessner, 1962). Some years after the discharge measurements had been made, computations based on 1972–1973 observations of isotopic concentrations (Oxygen-18) were also used to assess the proportion of water delivered by the two confluents. Here again, calculated results are broadly compatible with direct measurements. According to the ratios of isotope abundance, the relative share of the Negro was greatest in July 1972, with 43± 9 percent and smallest in January 1973, with 20±5 percent (Matsui *et al.*, 1972).

The fact that optical differences are no longer perceived several tens of kilometers downstream does not signify that a complete mingling has been attained. Thus, for instance, some 100 km below the confluence the amount of dissolved salts, as measured by electric conductivity, still shows significant variation, increasing from the left to the right bank (Gessner, 1962) (*Fig. 24*). Similarly, even at a distance of 120 km, differences

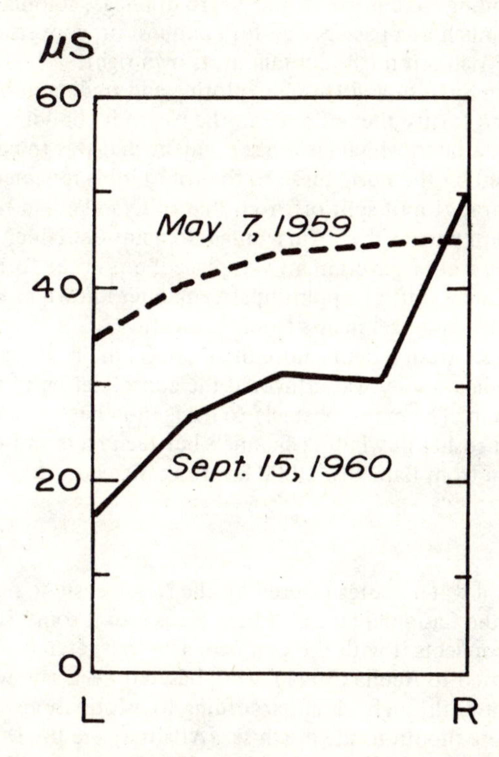

Fig. 24. Amount of dissolved salts in the Amazonas, as expressed by the electrical conductivity of the waters. Surface samples at Amatari, about 100 km downstream from the junction of the Negro and Solimões. L, left bank; R, right bank. (Gessner, 1962).

in the ratios of isotope abundance in the cross-section showed that a thorough lateral mixing of the two confluents had not occurred (Matsui *et al.*, 1972).

Whereas ongoing investigations are identifying and quantifying dissimilarities in water quality, some of the ecological repercussions thereof have long been recognized. Thus, Indians and Portuguese had acquired practical knowledge of the low level of biological productivity of the Negro's waters. It was noted, *inter alios,* by Wallace (1853): "no islands of floating grass, no logs and uprooted trees . . . scarcely any stream, and few signs of life in the black and sluggish waters."

If inhabitants or travellers along the Negro have lamented the scarcity of victuals in this "river of hunger," they have also recorded a relative freedom from certain insect pests. Some students of plant-derived insect hormones and hormonal analogues have speculated that the Rio Negro "must be full of insecticide" (Williams *et al.,* 1967), more particularly, that it may contain large concentrations of substances evolved in plants as a defense against phytophagous predators, and having endocrine action on insects (Kafatos *et al.,* 1967). The significance of such a finding, if confirmed, need hardly be emphasized: hormonal compounds, being specific to insects in their effects, hold forth the promise of a new, environmentally sound method of pest control (Bowers, 1971).

The thought that the Rio Negro, as "the world's most abundant plant extract," may contain in itself a valuable resource represents quite a novel viewpoint. Indeed, one suspects that, in contending for control of the Negro drainage, colonial powers for centuries saw it at least as much as a passageway for conquest or, conversely, a backdoor to be defended, as for any riches it might contain in its own right.

Sensing a threat in Spanish and Dutch exploring and trading parties, the Portuguese decided in 1669 to fortify the entrance to the Negro by building the fort of São José do Rio Negro, around which grew a settlement that was to become Manaus. The importance attached by the Portuguese to the tributary is reflected in the name given to a new administrative unit split off from that of Grão Pará in the late 1750's to encompass the western part of their Amazonian domain: Captaincy of São José do Rio Negro. Initially, the seat of government was placed, not in the fortified settlement at the mouth of the affluent, but at a point upstream, later known as Barcelos, more advanced vis-à-vis possible encroachments from the north.

Despite the distance from Belém (more than 2,000 km, in the case of Barcelos), relative ease of communication by water favored the consolidation of the Portuguese hold on the Negro system. The major obstacle to navigation begins at São Gabriel da Cachoeira, where water rushes in winding channels between reefs and rocky islands, about 500 km upstream from Barcelos, 1,000 km from Manaus.

The Casiquiare Canal

In terms of natural water routes offered by the Negro system, none has received greater attention than the Casiquiare canal, which leads into it some 300 km above the São Gabriel Falls and connects it with the Orinoco. The first reference to this connection is generally attributed to Acuña (1641), who, based on hearsay, wrote of "an arm thrown off by [the Negro], through which, according to informations, one emerges on the Rio Grande, at whose mouth in the north sea [Atlantic], are the Dutch." Acuña, it is worth noting, expressly excluded the possibility that this river might be the Orinoco. The earliest first-hand description of the Casiquiare recorded in scientific literature appears to have been that of a Spanish Jesuit, who, journeying up the Orinoco, encountered a Portuguese slaving party from the Rio Negro, fell in with it as it returned home through the Casiquiare, and later made his way back to the Orinoco by the same route. The account was divulged in 1745 by La Condamine (1749) during a session of the French Academy of Sciences.

Many were incredulous regarding the existence of the canal, and Humboldt (1822), who travelled through the 355-km long Casiquiare in 1800, appears to have derived particular satisfaction from the fact that his expedition had completely refuted the doubts raised by the eminent geographer Philippe Buache, who scoffed at the interfluvial linkage as a figment, a "*monstruosité en géographie.*"

That the Casiquiare carries off a considerable fraction of upper Orinoco water (estimates vary from 1/8 to 1/4, depending on river stage), suggests an ongoing process of stream diversion (*Fig. 25*). In fact, the present drainage pattern has been regarded by some as a "fine example of unfinished river piracy" (López, 1956; see also Eden, 1971). Other authors deny that the Casiquiare is "the result of capture by one stream of another" (Rice, 1921), and object to the implication that the shifting of the divide was effected by headward erosion of the Negro system, being persuaded that the Orinoco "took the entire initiative of the connection, periodically spilling its floodwaters over a low point in the divide" (Paiva, 1929).

The role of "small crustal dislocations" has been mentioned in connection with the diversion (Rice, 1921), but the hypothesis appears to have met with little favor, being considered "unnecessary and difficult to corroborate" (Paiva, 1929). Subsequently, however, considerable evidence has been published pointing to the importance of tectonic activity in Amazônia. In this respect, attention is called to a second communication between the Negro and the Orinoco: the Maturacá, which links the Cauaburi, a tributary of the Negro, to the Baría, draining into the Casiquiare (see location map in figure 25). It was described and mapped in the 18th century (Almada, 1786) and apparently used by balata and rubber traders during the early 20th century (Paiva, 1929). The entire route through this "miniature Casiquiare" (Sousa, 1959) may well lie under 100 m, but, towering above it, less than 50 km away, is Brazil's highest mountain, the Pico da Neblina. which exceeds 3,000 m (*Fig. 1*). Such a steep gradient suggests faulting; one cannot exclude the possibility that crustal deformation may be one of the determinants in the origin of the Casiquiare and similar interconnections.

The fact that palynological and other data from northern South America are being interpreted to indicate periods of very dry climate (see, for instance, the pollen studies of Wijmstra and Van der Hammen, 1966) suggests another possible line of investigation: that the derangement of the divide between the Orinoco and the Amazon systems may have been favored by a sequence of alternating arid and humid climates, such as is hypothesized for the Caroni River in Venezuela (Garner, 1966, 1967).

Whatever factors may have contributed to the origin of the Casiquiare --and the necessary geomorphic investigation remains to be done--, unless the present trend is reversed, the Orinoco above the Bifurcación, in due course, is likely to be entirely diverted in favor of the Negro.

The importance of the Casiquiare passageway is immemorial. Travellers have long noted the "prodigious . . . variety of idioms" (Humboldt, 1820) spoken in the Orinoco, Casiquiare and Negro region. Here lies what seems to have been one of the most significant cultural thoroughfares in South America (Nordenskjöld, 1916). Indeed, there is some suspicion that the Casiquiare route played an important role in the northward diffusion of Tapajo-style pottery (Lathrap, 1970), which shares common traits with ceramics

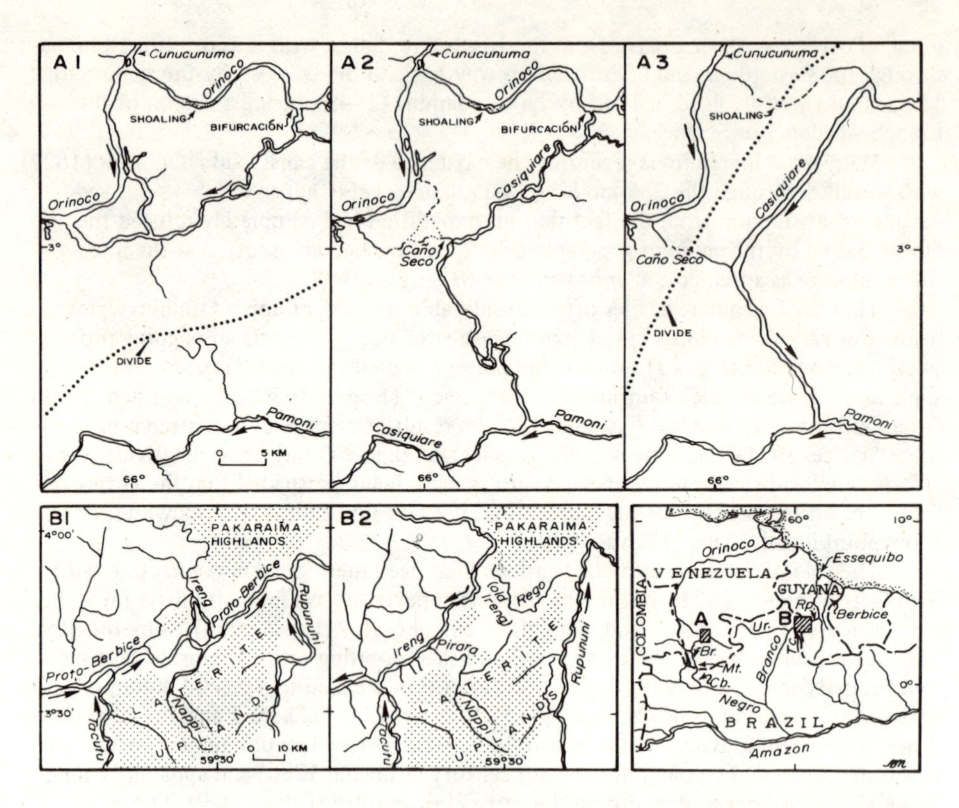

Fig. 25. Hypothetical development of drainage along the northern divide of the Amazon Basin. *A.* – The Casiquiare diverting water from the upper Orinoco into the Rio Negro – past, present and future. According to Vareschi, 1963, and Stern, 1970; the interpretation is that of an unfinished river capture by headward erosion of the Rio Negro drainage. *B.* – The Pirara depression, visualized as the former channel of an eastward-flowing drainage network (the Proto-Berbice River) captured by the Rio Branco. According to Sinha, 1968. The idea of simple river piracy has been disputed in these fluvial interconnections and possible factors involved in the derangement of the divides are mentioned in the text. Abbr. in location map: Br, Baria; Cb, Cauaburi; Mt, Maturacá; Rp, Rupununi; Tc, Tacutu; Ur, Uraricuera.

from the Orinoco delta and even the southern part of Central America (Nimuendajú, 1949).

Although not a shred of evidence has been found to give substance to speculations that the Casiquiare itself might have been partially dug by pre-Columbian man (Koch-Grünberg, ND), a considerable amount of thought has gone into how his modern successor may improve the channel or even construct alternates for it. Humboldt proposed to "facilitate . . . the communications between . . . Spanish Orinoco and the Portuguese possessions on the Amazon" by means of an artificial canal that would bypass the "Casiquiare, full of sinuosities and feared for the force of its current" (Humboldt, 1822). During the second world war, a study was made by the US Army Corps of Engineers

(1943) with the objective of developing the Casiquiare as an inland route that would avoid German submarines in the Atlantic. The report proposed several plans, which included the improvement of existing river channels by blasting and dredging, and the canalization of rapids, with the aid of dams and locks.

The most recent proposal for an all-water route between the Orinoco delta and the Rio Negro, as the first stage of a continental system of navigation, utilizing the Orinoco, Negro, Amazon, Madeira, Guaporé, Paraguay and La Plata (Curiel, 1971; Venezuela . . ., 1972/73), was outlined by the Minister of Public Works of Venezuela during a visit to Washington in 1971. At the request of the Ministry, the Inter-American Development Bank helped make a preliminary--and thus far, confidential--survey of the proposed project. In 1974, the Minister announced that Venezuela had requested United States technical assistance, in order to accelerate the implementation of the proposed interconnection. One feasibility study for a waterway such as that desired by the Venezuelan government entails the removal of some 92 million m^3 of rock and requires the digging of an artificial channel through a 26-km granite divide. The excavation techniques described involve high-energy chemical and nuclear explosions, as well as conventional dredging operations (Paz-Castillo and Kruger, 1972). Although such a project cannot fail to fascinate the world's development-mongers, it, like the long-envisioned canal between the Amazon and Plata basins (Mattos, 1949; Brasil, 1968; SGTE-LASA, 1971), is fraught with ecological hazards not necessarily limited to the more obvious ones associated with nuclear excavation.

The Rio Branco

Acuña (1641) was informed, at the mouth of the Negro, that the "lands of this river" contained "many and good *campiñas* covered with palatable pastures to graze innumerable heads of livestock," a description that could only refer to the grasslands lying some 900 km by river to the north, within the watershed of the main subaffluent, the Branco.

Most of the Branco system is inscribed on an extensive, low-lying surface, cut on the gneisses, quartzites and granites of the Guiana shield and interrupted by occasional bosses and ranges. The lower course of the Branco flows through a narrow alluvial valley set into this crystalline basement. Subject to the contingencies of great seasonal fluctuations in stage, navigation by small river boats is possible as far as Caracaraí, some 400 km from the mouth, where a stretch of rapids of more than 20 km marks the terminus of the lower course, and signals that the river here is cutting into the oldland.

The rapids can be negotiated by small craft at high water, and a short portage trail has long been used for year-round communication with the Alto Rio Branco, where the equatorial rain forest gives way to campos, estimated to cover 40 to 50 thousand km^2 in Brazil and an additional 12,000 km^2 or so in Guyana: the Rio Branco-Rupununi savannas. The northward rise of the bevelled shield traversed by the river is gradual. At the confluence of the Uraricuera and Tacutu, in the very heart of the Alto Rio Branco region, riverine lands lie less than 100 m above sea level. Throughout most of this savanna-covered area, the crystalline rocks are buried under a mantle of unconsolidated, often clayey, sands (with evidence of a graben underlying the Tacutu River, *cf.* McConnell *et al.*, 1969). The fluvial network is poorly organized in the gently undulating terrain and, at the height of the rainy season, inadequate to evacuate the runoff. Streams top their banks, and vast sheets of water wash over the borders of hydrographic systems and subsystems.

To the east, spreading over a broad lowland commonly named after the Pirara River (a subaffluent of the Tacutu), the annual overflow erases the divide between the Rio Branco system and that of the northward-flowing Essequibo River, in Guyana. Referring to a passage between the two basins by way of their respective tributaries, the Tacutu and Rupununi, a reliable 18th-century source reports that it constituted an "ancient communication" of the Rio Negro Indians with the colonies of Dutch Guiana (Sampaio, 1825).

At this point, one may reexamine Acuña's reference to a route between the drainage of the Negro and the "Grande," an unidentified river having Dutch settlements at its mouth. Although Acuña's description generally is construed to mean the Casiquiare, such an interpretation does violence to the priest's singular insistence that he could "definitely assert that the Grande, in no way, is the Orinoco." The rumored connection alluded to by Acuña may not have been serviceable throughout the year. As Humboldt, for one, has pointed out, "periodical innundations, and above all *portages* by means of which canoes are passed from one affluent to another . . . lead one to assume non-existent bifurcations and branchings of rivers" (Humboldt, 1822). Since Acuña appears to have picked up a reference to the savannas of the Alto Rio Branco, it is conceivable that the passage he heard about, linking the Negro and an ill-defined Rio Grande, may not have been the Casiquiare at all, but the Rio Branco-Pirara lowland. In this case, the unidentified Grande really would be the Essequibo, which, according to Kemys (1596), the Indians "to shew [its] worthiness . . . do call it the brother of the Orenoque" and which, as a matter of fact, had been settled by colonists from Zeeland at least since 1616, a quarter century before Acuña's journey.

The seasonally flooded lowland between the Branco and the Essequibo drainage is commonly interpreted to have resulted from stream diversions in favor of the Amazon drainage, as in the case of the Casiquiare, albeit at a more advanced stage. The right-angled turn of the Uraricuera and the crook of the Tacutu, both near their confluence, are held to be of diagnostic value in hypothesizing stream piracy (*Fig. 26*). If such bends are indeed "elbows of capture," they mark the points where the two rivers, once tributary (by way of the Rupununi) to the Essequibo (Passarge, 1931; Barbosa and Ramos, 1959) or to an even more easterly "proto-Berbice" (Sinha, 1968), were intercepted by headward erosion of the Branco (*Fig. 25*). According to this explanation, the Pirara lowland is the site of the former channel that led off to the east the waters of the Uraricuera, the Tacutu and its affluent, the Ireng. Concerning factors that might have been involved in the presumed migration of the divide, the reader is referred to what has been said in the matter of the Casiquiare. Since old structural trends were reactivated before, during the creation of the Tacutu rift, it is reasonable to speculate upon the possible role of crustal mobility in the recent rearrangement of the hydrography.

During the rains the lowland is converted into a lake. Despite the seasonality of this feature, word of the inland body of water, relayed down to the north coast, was consistent with the persuasion that a "great Golden City which the Spanyards call *El Dorado* and the naturals *Manoa*" (Ralegh, 1596) lay on the shores of an interior sea between the Amazon and Orinoco. Belief in the "*Lago Dorado* that maintains so restless the spirit of everybody in Peru" (Acuña, 1641), was widespread, and considerable resources were mobilized for the treasure hunt. Francisco Xavier Ribeiro de Sampaio, a shrewd Portuguese official, writing in the 1770's, suggested that the Spaniards' stubborn quest might even be called "the philosopher's stone of the discoveries" (Sampaio, 1825).

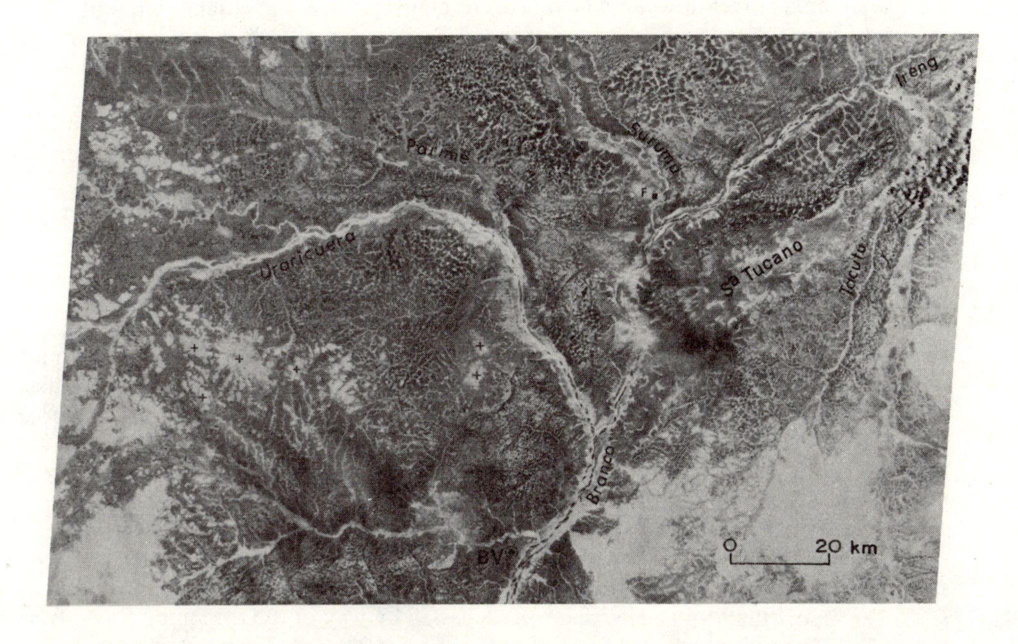

Fig. 26. Headstreams of the Rio Branco. With the exception of crystalline basement rocks surfacing here and there (*e. g.* areas marked +), consolidated sedimentaries of the Tucano block and isolated volcanics, most of the area shown is one of clays, sands and different types of lateritic materials. The hydrographic filigree of the low-lying savanna lands, with closed lakes and interconnecting streams, shows lack of integration. Vast sheets of water spread out in the flood season, and, by way of the Pirara, a tributary of the Ireng (or Maú), give access to the northward-flowing Rupununi (Essequibo basin). Indeed, some students believe that, before being diverted by the Branco, the waters of the Uraricuera, the Tacutu and their tributaries drained regularly eastward. With respect to this hypothesis, note the arcuate depression between the Parimé and the Tacutu. (NASA ERTS-1 Image, 4 March 1973).

Cartographers for a long time sanctioned the existence of the lake, whose legendary treasures fomented the covetousness of governments and lured countless adventurers from England, Spain and other countries of Europe. The Dutch, "as imaginary neighbours" (Sampaio, 1825) of the legendary lake, had noted that "in the River Essequibo there is situated a river or water which takes its course inland, and extends without interruption to the Lake of Parima, or Rupowini as it is named, which is very famed for its richness in gold, and near which Lake of Parima also not far off is situated the so-called place Manoa El Dorado," (West India Company, 1714). Accordingly, the chartered West India Company directed the Commander of Essequibo "to proceed up the aforesaid river . . . and then to find out whether possession might not be obtained at the aforenamed places." As a matter of fact, the first European recorded to have penetrated the Rio Branco coming from the Essequibo, by way of the Pirara lowland, was a German in the service of the Dutch, who, "after traversing lakes and vast plains, now dragging, now carrying his canoe," had arrived about 1740 at the Rio Branco (La Condamine, 1749).

The Portuguese, who indulged in other chimeras (such as the elusive "Serra das Esmeraldas"), appear to have been less aroused by the legend of Lake Parime (or Parima) than their neighbors. Sampaio is worldly-wise: El Dorado "should only be treated in the allegorical and ironic manner in which a famous author has written of it." The reference is to Voltaire, whose satirical *Candide* in 1759 contained a deliberate caricature of the imperial city of Guiana. Yet, while recognizing that Lake Dorado exists "only in the imagination," Sampaio and his countrymen maintained an unrelaxed interest in the very tangible consequences of the exploratory operations it generated. Whatever the fanciful alchemy of a rival's territorial aspirations, these were never taken lightly. They constituted geopolitical realities requiring counteraction.

The report that "some Hollanders have passed by way of the Essequibo River, from the lands of Surinam to the Rio Branco" led in the 1750's to the building of a fort at the mouth of the Tacutu. The fact that Dutch slave-hunting parties, operating out of the Guianas, had sallied into the Rio Negro drainage lent urgency to the proposed establishment of a Portuguese settlement on the banks of the Rio Branco, "where there are excellent grasslands and good airs." By royal command, the riverine lands were to "be patrolled, from time to time and *especially during high water*" (my italics) and some villages were to be "established in those confines" (*Carta Regia* . . ., 1753).

The Captaincy of São José do Rio Negro was to extend on the north and west to the Spanish domains. Portuguese strategy, pegged to the Amazon network, dictated the "indispensable need to people this western frontier and, in securing it, to secure the navigation of the Madeira to Mato Grosso and the passage from those mines to the Cuiabá" (Pombal, 1755). After all, even if there is not an outright mouth-to-mouth contraposition between the Negro and the Madeira, the intervening distance is not great and is further reduced by floodplain channels leading from the Solimões to the Madeira.

As to the recommended peopling, insofar as the Rio Branco was concerned, a viable basis was seen in the "vast grasslands of that river [that] are calling for livestock" (Ferreira, 1786). Cattle would provide new sources of revenue--as well as food. The first cattle were introduced in 1788, three ranches being established at the junction of the Uraricuera and Tacutu. Even the commander of the blockhouse on the Tacutu, enthusiastic over the prospects of the undertaking, threw himself into it and ordered twelve heifers (Reis, 1940). Thus was launched the beef-cattle industry that, with a herd numbering some 240,000 head (1970), to this day dominates the sluggish economy of the

Fig. 27. Shipping cattle in the Rio Branco drainage (Roraima Territory). Not much has changed in the operations since 1948, when this photograph was taken at the Fazenda Flexal (right bank of Surumu River, *F* in Fig. 26). The often semi-feral cattle are rounded up during the high water season, driven to a riverside corral and, by means of a chute, packed into cattle boats. In this case, a hundred head were being shipped to Manaus on two barges secured on either side of a motor launch. A deck above the cattle, on one of the barges, served as living quarters for crew and passengers. (H. O'R. S.).

Rio Branco. Pastures have not corresponded to the promise of excellence read into them, and zootechny has made slow headway. Up to very recently, stockmen have done little more than round up semi-feral scrub cattle, to be sold locally or loaded on crowded barges for an exhausting trip downriver to Manaus (*Fig. 27*). Laborious at best, this form of shipment becomes well-nigh impossible during the dry season.

Cut off from the remainder of Brazil during low water, except for air transport, *riobranquenses* for decades have been demanding an all-weather road to connect their Territory with Manaus and the world outside. Such a linkage is now being pushed through to Boa Vista, the territorial capital and, with a population of 16,720 (1970), the only town of any significance. The Manaus-Boa Vista road, part of the Panamerican system, eventually will tie in with the Venezuelan highway network at Santa Elena de Uiarén. The stretch from Caracaraí to Boa Vista already has been built, obviating the need for transhipment to the upper course of the Branco. The road has given a great boost to the little cluster of houses at Caracaraí, seat of a predominantly forest-covered *município* of 133,603 km^2. The nucleus started as an encampment of ranch hands at the point where cattle in transit from Boa Vista were reembarked, after bypassing the rapids on the hoof. The first bunkhouses date from the turn of the century, but most of the buildings now in existence were put up after the advent of the road.

Although recently there has been a rapid increase in the "urban" population of the old cowboy camp (estimated to have passed 1,000 by 1973), this growth will be as noth-

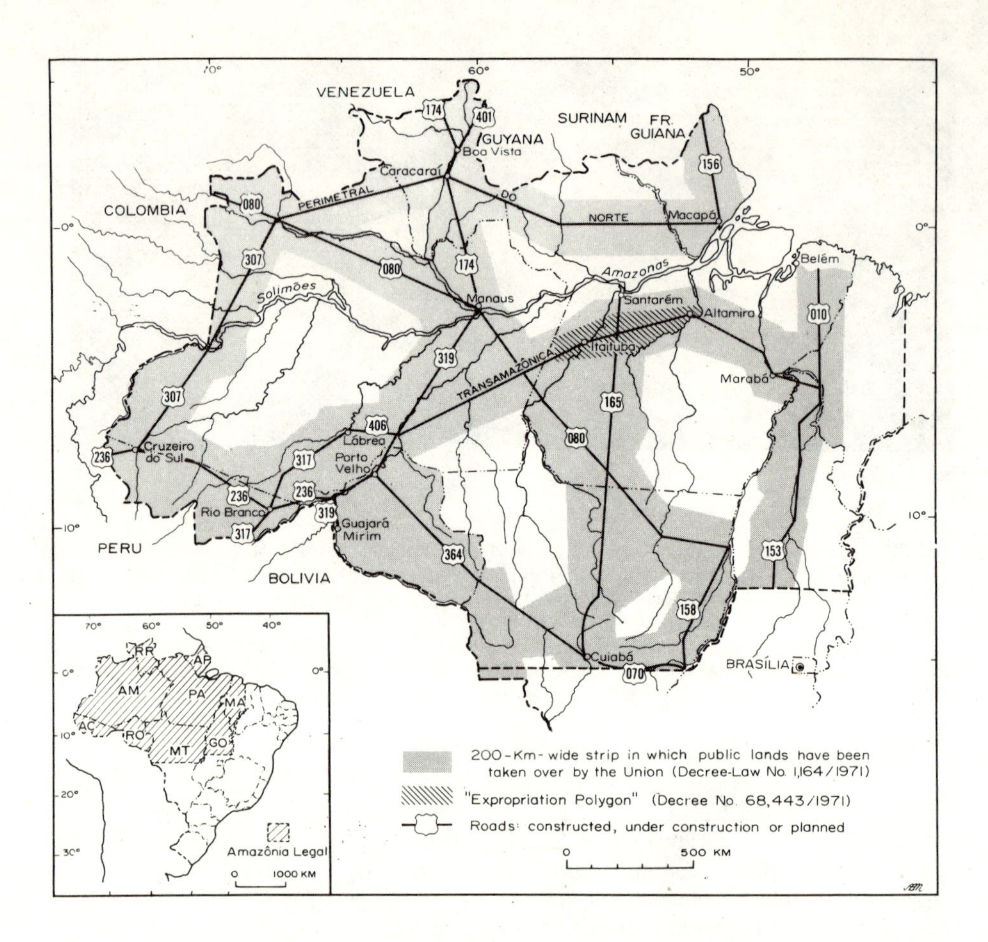

Fig. 28. Projected Highway Network in Amazônia (INCRA, ND). Ten years after the Belém-Brasília highway (BR–153 and BR–010) was symbolically opened to traffic, Decree-Law No. 1,106/1970 established a Program for National Integration, to be promoted by the construction of the Transamazon and Cuiabá-Santarém highways. To expedite resettlement in Amazônia of rural folk, mostly from the northeast, a 1971 Decree opened the way for expropriation of an area of some 65,000 km^2 along the Altamira-Itaituba section of the Transamazon. Almost simultaneously, the federal government took over from the states control of all public lands within a 100 km strip on each side of the Amazonian highways. An official of the National Institute for Colonization and Agrarian Reform, presumably having subtracted Indian and titled lands from the 3 1/2 million km^2 affected by the legislation, estimates that almost 2 1/4 million km^2 of real estate will remain available for development, mainly as farms and ranches (Stephanes, 1972) Abbr. in inset map for states and territories entirely or partly within Amazônia Legal (some 5 million km^2, or roughly 60 percent of the national territory): AC, Acre; AM, Amazonas; AP, Amapá; GO, Goiás; MA, Maranhão; MT, Mato Grosso; RO, Rondônia; RR, Roraima.

ing compared with the expansion that is projected for Caracaraí. Indeed, this settlement marks the spot where Brazil's more than 4,000-km long Perimetral do Norte road, under construction since July 1973, eventually will intersect the Manaus-Boa Vista highway. The first leg (about 2,500 km) of the perimetric or rim road, starting from a point near Macapá, thrusts westward, curving north to Caracaraí--which may become northern Amazônia's major transportation node--and, passing São Gabriel da Cachoeira on the Rio Negro, terminates on the Colombian border, not far from the equator (*Fig. 28*).

RIVERWAYS AND ROADWAYS

WATERBORNE TRANSPORT

The Perimetral, like the Transamazon, is part of a gigantic highway mesh now being superimposed on the Amazon River system, which is estimated to offer about 20,000 km of navigable channels, much more in the rainy season.

Potentialities for navigation of such waterways vary throughout the Amazon network. The main river really constitutes a unique extension of the sea, with the head of deep-draft ocean navigation at Iquitos in Peru, more than 3,700km from the coast (SUDAM, 1971a). Favorable conditions extend for some distance up the drowned estuaries of such affluents as the Negro and the Tapajós, whereas white-water tributaries, although in some cases allowing access to relatively large river steamers, are not without problems, in terms of floating or entangled logs, hidden snags, slumping banks, changing channels-- or, simply, time-consuming sinuosities (*Fig. 13*). Thus the Indian name of the Tarauacá, an affluent of the upper Juruá, is rendered as "river of tree trunks" (Cashinawa: *tará,* trunk; *waká,* river) by Tastevin (1925), who describes (1926) one section where the log jam was such that "at first glance one would judge the passage to be impossible." The middle and upper courses of tributaries cascading from areas of marked rainfall seasonality offer at best precarious, often intermittent, transportation. As an example of extreme variation in stage, one may cite the waters of the Acre River, which are reported to rise and fall as much as 25 m at Rio Branco, capital of Acre state (Bischoff 1963b). Furthermore, lack of overland connections between the tributaries of the Amazon can be a major problem. Thus, to go by launch from Rio Branco to Cruzeiro do Sul (1 hr. 45 min., by air) one must travel down the Purus, then up the Solimões and the Juruá-- a distance of the order of 5 1/2 to 6 thousand kilometers-- easily a month's trip.

Whereas the natural navigational capabilities of the Amazon system are not so great as might be assumed from looking at a map, it is only fair to say that the very considerable assets of geography have not been maximized and, indeed, in some cases have been cancelled out by a skein of man-made encumbrances.

Docking facilities upriver are non-existent, except at Manaus; even relatively significant river towns, such as Itacoatiara, Parintins, Humaitá and Porto Velho have only rudimentary terminals (the same has been true of the third-ranking town of Brazilian Amazônia, Santarém, where new wharves are now under construction). Appropriate craft adapted to the conditions of the region are also lacking. A survey of Amazon river navigation carried out almost a quarter of a century ago reported the "impressive deterioration of mostly obsolete equipment" (SNAPP, 1951). Hardly enough has been done in the intervening years to reverse the trend and develop a forward-looking waterways plan. Indeed, for a part of this period, institutional distortions introduced by populist governments resulted in the mandatory hiring of superfluous crew members and the granting of extravagant concessions to shipside labor (Sternberg, 1965). In view of the high cost of inefficient operations and rampant pilferage, shippers were ready to forsake the waterways in favor of the quicker, safer and, all things considered, cheaper transportation by truck, as soon as this alternative became available.

"DEVELOPMENT TYPE" ROADS

Current plans for Brazilian Amazônia call for a reticulum of 17,855 km of trunk roads (*Fig. 28*). Indeed, there has been over the last two decades or so a considerable change of heart in the relative importance attributed to different transport modes for Brazilian Amazônia. In 1955, according to the agency then responsible for developing the region, fluvial transportation was the only one fit to cope with the "cyclopean" distances without adding excessively to the cost of the freight carried (SPVEA, 1955). More recently, the waterways have been judged capable of contributing only in a supplementary manner to the top priority objective of national integration (SUDAM, 1971a). The advantages that water haulage offer for certain types of freight have been overridden by developmentist impatience. The authors of one study of the Belém-Brasília highway are unequivocal: "Those who opposed it *or defended the system of mixed communications (rail-fluvial or highway-fluvial)* (my italics) did so, in most cases, for lack of vision of the problem or exaggerated conservatism characteristic of the 'colonial mentality', to which N. Werneck Sodré refers in *Ideologia do Colonialismo*" (Valverde and Dias, 1967).

In weighing the economic benefits expected to accrue from the Amazon highway network against its cost, policy-makers are operating within a rather special frame of reference. Take, for instance, a feasibility study for one of the links in the Amazon mesh, the approximately 866-km long Manaus-Porto Velho highway (BR-319), whose roadbed already is in place and which follows in part the Purus-Madeira watershed (*Figs. 13* and *29*). The report concludes that "Considering only benefits from normal, generated and diverted traffic [*i. e.*, diverted from the waterways], the road is shown to be economically unfeasible" (TRANSCON/Berger, 1968). However, like most national and international highways in the Amazon basin, such as Venezuela intends to build for "La Conquista del Sur," or Peru and Colombia designate as the "Carretera Marginal de la Selva," this link is regarded as a "development type road" (TRANSCON/Berger, 1968), whose objective is to "open up vast areas of one of the world's last frontiers" (Stokes, 1966).

Both the Transamazon (*Fig. 30*) and the Perimetral highways roughly parallel the Amazon. The former crosses major right-bank affluents and subaffluents approximately at the head of uninterrupted navigation, generally several hundred kilometers from the Amazon. The latter lies between this river and the boundaries of Brazil with Colombia, Venezuela and the Guianas; most of the relatively short left-flank tributaries draining the Guiana shield are interrupted by a succession of falls and rapids that begins on resistant strata of the Paleozoic (Bischoff, 1963a), quite near their mouths (*Fig. 31*). The first leg of the Perimetral therefore will open up country that has been relatively inaccessible; in fact, it has been a refuge for, according to one estimate (Brasil, 1973), about 30,000 Indians, half of whom remain in complete isolation--some of the last relicts of the millions of pre-contact Amazonians. The effectiveness of such highways in economic terms will be judged by, for instance, the value of resources whose discovery and exploration they make possible in hitherto inaccessible interfluves.

Fish and other biological resources of the rivers themselves are to be tapped and drained off by means of the new highway network. Thus it has been deemed advantageous that the Transamazon will put "the well-stocked southern tributaries of the Amazon in contact with the Northeast with its more than 30 million inhabitants who lack animal protein," and that the market will be further expanded by extending the roads into Bolivia, Peru, Ecuador and Colombia (Azevedo, 1970). Now, whereas fish and turtle *farming* would seem to offer great opportunities in the Amazon, there is evidence that

Fig. 29. Construction of Manaus-Porto Velho highway (in 1970). The northern end of the road begins after a stretch of water of approximately 10 km, to be crossed by ferry, between the Manaus uplands and the right bank of the Paraná do Careiro. An extensive section of the Careiro-Humaitá leg cuts through floodplain forests and is being constructed on embankments, requiring a great number of culverts and bridges. (H. O'R. S.).

Fig. 30. The decision to build the Transamazon highway (not included in the 1967 federal transportation plan) was put into force by means of a 1970 decree, creating a National Plan of Integration. The road was located on the sometimes surprisingly hilly uplands, so as to link the heads of navigation of right bank tributaries of the Amazon. Whereas some of the road-building equipment could be trucked in, the heavier machinery initially was brought in by water. (A. Tamer, courtesy *O Estado de São Paulo*).

the riverine fauna does not bear up well in the face of high-powered exploitation. Recent denunciations have focused on the fishing-out of streams in Goiás and Mato Grosso as the result of predatory commercialism. The complaints brought out the fact that refrigerated trucks, carrying fish from as far away as the Guaporé and Acre regions, were passing through Cuiabá, on their way to outlets in São Paulo and other west-central or eastern cities (*Memorial* . . ., 1972).

GEOPOLITICS AND TRANSPORTATION IN AMAZÔNIA

Actually, it would be quite unrealistic to discuss in purely economic (not to speak of ecologic) terms those road-building programs that impinge upon the Amazon basin. Countries that share this, the greatest hydrographic network in the world, have not yet effectively integrated the Amazonian lands into their body politic. International borders lie in areas of low demographic and economic density, "inadequately" linked to their re-

Fig. 31. Falls on Jari River. Like other tributaries that flow from the Brazil-Guianas divide, the Jari, after a short navigable stretch near the mouth (about 100 km), is interrupted by rapids and falls. Jules Crevaux (1883), after the hardships of an exploration of the Jari, reflected that it was "not without reason that the Indians of the Paru and the Jari traverse the [Tumuc-Humac] mountain to barter in the Oyapock, rather than descend their rivers." Rubber tappers, Brazil-nut gatherers and, occasionally, prospectors are the only travellers in the upstream areas, and Belém-based grubstakers, strategically locate company stores along the "fall-line". (H. O'R. S.).

spective political nerve-centers. The long tradition of applying the principle of *uti possidetis* to resolve boundary questions between the Iberian colonies and, later, between the countries arising from them; previous experience with infiltration along borderlands; past alarums with the incorporation abroad of trading companies or agri-businesses destined to enjoy full powers of states-within-states (see, for instance, Reis 1960) --all seem to have reinforced in South America a concern with the occupation of such remote portions of the national territory. Apprehensions may lie dormant for a time, but, as former United States ambassador to the Soviet Union George F. Kennan once pointed out in a different context, "Every government is in some respects a problem for every other government." A program of settlement in Amazonian areas conceived in terms of highway construction and embarked upon by one country sets up reverberations among its neighbors: perception of the frontier as a liability for national sovereignty reaches a threshold where "symmetrical" action is called for. The felling of trees for a pioneering highway eventually produces a domino effect in the forest across the border.

Pronouncements are many that equate colonization of Amazonlands with national security. Take Peru, 1965: the preamble of a "Supreme Decree" establishing the program for colonization in the forests of the country's borderlands points out that "the greater part of the boundary zones of the selva are uninhabited and insufficiently developed, constituting an obstacle for the full dominion of national sovereignty and the better utilization of the natural resources of the country." The decree recognizes the army as principal executor of the program (*El Peruano,* 1965). Peru, 1974: the newsweekly *Oiga,* under the title "The Army takes the Revolution to the Selva", describes "an accelerated and planified colonization [in Amazonía], which, because of the military character the task assumes, significs that, since 1972 an acute border consciousness has emerged at a vertiginous rate. In practice, the Peruvian plan, according to the commentator of a news agency, seems to be an answer to Brazil's march toward Amazonía" (El Ejercito . . ., 1974).

In short, all projects in the Amazon basin, especially along international boundaries, tend to be equated in terms of political-strategical concerns, in relation to which final decisions will always rest with the military. Brazil is no exception (SUDAM, 1967).

THE PROGRAM FOR NATIONAL INTEGRATION

A narrow definition of strategy is not implied here, but rather one that covers the larger aspects of statecraft, with military and non-military factors interwoven in national policy. Thus, the immediate stimulus for actually launching the *Programa de Integração Nacional,* PIN, --of which the Transamazon and Cuiabá-Santarém highways are the cornerstone--was said to be the plight of the backlands population of northeastern Brazil, stricken by yet another drought in 1970. While the PIN stressed penetration and effective occupation of the vastness of Amazônia, the program also addressed itself to the source region of the settlers and was visualized as a means to "absorb the population of areas considered totally unfit for human life" (Médici, 1970). Indeed, the PIN also included the implementation of the first stages of an irrigation plan for the northeast.

With emphasis on the "peopling" of Amazônia as the result of relocation of farm families from the northeast along the Transamazon and the Cuiaba-Santarem axes, the decree creating the PIN reserved "for colonization and agrarian reform a belt of land up to 10 km-wide, left and right of the new highways." While heavy earth-moving equipment was ramming through these two pioneer roads, the federal highway plan was being updated to fit them into an ambitious Amazonian network. This includes the already-mentioned perimetric highway along the northern boundary, and several north-south as well as diagonal linkages, the whole binding together and bonding to the heartland of Brazil the operational region known as Amazônia Legal (*Fig. 28*).

The policy heralded by the enactment of the PIN, with its reference to agrarian reform, was one that favored the family farm; it did so with the dual objective of settling Amazônia and of acting as a safety valve for the land-hungry northeasterners. Consistent with this approach, a follow-up decree, promulgated less than a year later (Brazil, 1971a), declared all privately owned rural lands within an area of some 65,000 km^2, straddling the towns of Altamira and Itaituba, to be "of social interest for the purpose of expropriation" (*Fig. 28*). This so-called "expropriation polygon," containing patches of some of the better upland soils known in Amazônia, was to be the setting for the first agricultural colonization project based on the distribution of 100-ha lots and the establishment of a supporting urban hierarchy.

At the same time, a Decree-Law proclaimed indispensable for national defense and development all *terras devolutas* (public lands) within a 100 km-wide belt on each side of the almost 18,000 km of Amazonian highways, then constructed, under construction or planned (Brazil, 1971b). In comandeering the terras devolutas, theretofore under the jurisdiction of the states, the federal government routinely pledged respect for Indian rights (*Fig. 32*), and held out the prospect that squatters, through habitation and cultivation, might legitimate their holdings on public lands. An official of the National Institute for Colonization and Agrarian Reform, presumably having subtracted Indian and titled lands from the 3 1/2 million km^2 affected by the legislation, estimates that almost 2 1/4 million km^2 of real estate will remain available (Stephanes 1972)--an area surpassed by no more than ten countries in the world. (*Fig. 28*). Settlement plans drawn up for the public lands within the 200-km wide strips, focused initially on the 10 kilometers immediately adjoining the highways and also emphasized the 100-ha farm.

The schedule for colonization attendant upon the 1970 decision to build the Transamazon called for the placement of some 100,000 families in Amazônia by 1975. It is now apparent that perfomance is likely to fall short of this target by a factor of 10 or more. Moreover, with the federal administration that took office early in 1974, emphasis swung away from settlement based on the family farm. The underprivileged tillers of the soil, who only yesterday occupied center stage, are now regarded with distinct coolness, at least by some key figures in government. Thus the Minister of Agriculture is reported to perceive no sense in allocating parcels of land to those "lacking technical or financial capability (*idoneidade*) to exploit them." Indeed, the official colonization policy that prevailed essentially up to the beginning of 1974 is being held responsible for promoting "the destruction of the forest and the exhaustion of the soil through the practice of subsistence agriculture, in the well-known system of shifting cultivation." The new line: "economic conquest of Amazônia by large firms." From the ecological viewpoint, one can only speculate why human sinew and ax should appear more destructive when used by the small settler than when backed up in the context of big enterprises with chainsaw, tractor and chemical defoliants sprayed by airplane (Almeida, 1974). Because of a misunderstanding prevalent in certain circles that equates the family farm with subsistence agriculture, it may not be amiss to point out in passing, that if the settlers man-

Fig. 32. Kreen-Akarore village in a region to be opened up by the Cuiabá-Santarém highway. For months, the agency for Indian affairs tried vainly to contact this tribe. The two Indians shooting at the airplane during one of the first overflights express the feelings of a group that consistenly has shown little desire to be integrated. The report of a 1972 fact-finding mission of the Aborigines Protection Society remarked prophetically that, if successful, attempts at pacification would put "an end to the independence of Brazil's most famous tribe." (Brooks, 1973). Indeed, in 1973, peaceful contact was established and, by mid-1974, according to the press, it was decided to use military planes to speed up the relocation of Kreen-Akarore in the Xingu National Park (Xingu tem . . ., 1974). (Anon.).

aged to produce no more than for their subsistence, or indeed practiced shifting cultivation in the short period of their tenancy, the fault lies not in the family farm system, but in the failure to establish it successfully. The same officeholder mentioned above, during a recent Senate hearing, stated that "it is impossible to transform every rural worker into an entrepreneur." There are students, however, who feel that the oft-decried lack of entrepreneurship in the Brazilian rural population can be ascribed, at least in part, precisely to a system of land tenure based on an unconscionable concentration of realty in the hands of a few. This mold already has been strengthened by the recent multiplication of stockbreeding firms in Amazônia, mostly organized in São Paulo for the purpose of tapping vast financial resources derived from tax incentives and administered by the Superintendence for the Development of Amazônia, SUDAM. The system will become even more entrenched with the promotion of corporate ranching along the new Amazonian highways. It would seem that what, rightly or wrongly, had been regarded as a social component in the overall strategy for settlement and integration of Amazônia has been displaced by what is now perceived by policy-makers as a more effective economic component.

By year's end 1973, SUDAM estimated that the 311 ranching projects it had approved for financial support were destined to engender no more than 14,549 jobs, despite the fact that they involved an area of 6,994,000 ha, *i. e.* about twice the size of Belgium. That much of the large-scale clearing of forest and grassification that lies behind the emerging land-intensive pattern represents an unjustified dilapidation of unknown and irreplaceable genetic resources has been pointed out elsewhere (Sternberg, 1973). But, considering the above ratio of jobs to space, one may wonder how the geopoliticians, ever fretting about "demographic voids," have weighed the family farm versus the large corporate ranch, in terms of the objective of "marking by the presence of the Brazilian man in Amazonian lands, the conquest for himself and for his Fatherland of that which was always theirs, so that no one may ever dare dispute them in this objective" (INCRA, 1972).

TRANSPORTATION AND THE ENERGY PROBLEM

Whatever may be the mix of social, economic and geopolitical considerations in the policy-making process, the Amazon Basin countries, like most developing nations, have misguidedly followed the transportation model of the United States, where six percent of the world's population consumes one third of the world's energy output. They have adopted the perspective of short-range economic cost, giving precedence to improved service, and favoring the most energy-intensive modes of transportation. With 1950 figures as a bench mark (=100), Brazilian indexes for the period 1950-1970 show that, nationwide, truck haulage (expressed in ton-kilometers and including the conveyance of such commodites as steel, cement, petroleum products and grain) increased more than thirteen fold during the two decades (1970 = 1,304.6). In the same period, other modes of freightage showed less than a four-fold increase (1970 = 340.9) (Barat, 1973). The expansion of highway transportation, entirely dependent on and a great consumer of petroleum, is in some degree reponsible for the fact that, despite increased domestic production, the country's imports of crude oil more than tripled between 1957 and 1970.

Brazil appeared to be in good company. In the United States, gross imports as a percentage of domestic petroleum demands had risen from 13.1 in 1950 to 35.7 in 1973 (Landsberg, 1974). Almost overnight, however, the apparently successful prototype, after a few generally unheeded signs of approaching trouble, was plunged into a disconcerting energy crisis. The ongoing shift in the balance of decision-making from petroleum-consuming to petroleum-exporting (or OPEC) countries and, notably, the Arab oil embargo, brought the problem to a head. But "the underlying causes lie further back in the past and hopes of long-term remedies lie well into the future" (Landsberg, 1974). Intermediate and long-range solutions will have to be sought within a framework of growing awareness that *all* forms of large-scale power generation now operative have a greater or lesser impact on the environment.

In the United States and its most "accomplished" emulators, a population reared to profligate spending, which has "traded energy for speed" (Hirst, 1973b), seems destined to be reeducated the hard way. It will not be easy to accept the fact that the quest for abundant and clean energy must go hand in hand with a policy of energy conservation, a change in entire patterns of transportation, *e. g.*, shunting freight, insofar as possible, from planes and trucks to pipelines, rails and waterways.

The very name, "Project Independence", given in the highly-developed United States to a national endeavour having as its 1980 target that of gearing the country to meet its "energy needs without dependence on any foreign . . . source," should furnish some food for thought to students and administrators in the developing world. Whether or not one believes that anything approaching energy self-sufficiency can be achieved in the United States, it is clear that the cause of the less-developed countries is ill-served by blue-prints for growth that unnecessarily impose an increased dependency on resources lying outside national borders. It is significant that, speaking at the first public hearing on Project Independence, in August 1974, the Federal Energy Administrator stressed conservation as the major way in which the United States can lessen dependency on foreign oil in the immediate future. "I don't want to hold out to the American people the hope of some Buck Rogers solution," he is reported to have told newsmen. "The focus will be on conservation."

The long-range implications of the global energy situation at first were dismissed rather lightly in Brazil. *Conjuntura Econômica,* published by the Brazilian Institute of Economics, reporting on the world crisis in its June 1973 issue, suggested that Brazil was "well prepared for the increase of international oil prices, provided the flow of supplies is maintained," (A Crise . . ., 1973). Six months later, the same periodical had adopted a more realistic perspective. It pointed out that about 75% of the total transport load in Brazil is carried over the country's highways and that 80% of railroad traffic is moved by diesel locomotives. The second article shows the extreme and increasing dependency on oil of Brazil's energy supply, and estimated that domestic output represented a mere 25% of the 40 million m^3 of petroleum products consumed in 1973. As now perceived, the quadrupling of the price of crude during that year accentuates the urgency with which the country should proceed to modify its policies concerning transportation, energy and location of economic activities, (Recursos energéticos . . ., 1974).

A recent and, in many respects, perceptive assessment of Brazil's transportation policies directs attention to the fact that simple extension of the road network and incorporation of new spaces into the economy traditionally have been preferred over improved operation of existing linkages and more intensive use of occupied areas. Such a choice translates an ambitious strategy of territorial occupation, but, says the author, the time has come to re-order the country's priorities and upgrade railroad and shipping technology (Barat, 1973). Although fluvial transportation is not mentioned, the case of Amazônia would seem to corroborate this thesis, which becomes even more cogent when seen through the optic of energy conservation and environmental protection.

Certainly, planners might well bear in mind the energy-intensiveness of different transport systems. Because of historical and geographical variations, energy consumption figures for inter-city transport in the United States during the mild-1960's are of limited transferability to Brazil and Amazônia in the mid-1970's. But, since they convey some idea of the relative energy efficiency of different modes of transport, they are given here, in joules per kg-km of freight hauled: pipeline, 330; waterway, 390; railroad, 490; truck, 1,700; airplane, 27,000 (Hirst, 1973a). A lopsided transport program for Amazônia, despite its avowed goal of "a perfect road-river interconnection," assigns relatively minor funds to the construction of high priority river ports and the improvement of waterways, while making heavy allocations for work on highways and airports (SUDAM, 1971b). One may anticipate that energy-intensive modes will be used increasingly for hauling, not only much low-grade tonnage presently carried by water, but also the new cargo of this kind generated by the opening-up of the region. A glance at the freight trucked along the slightly more than decade-old "Belém-Brasília" highway is suggestive, since it includes many commodities that, in the interest of energy conservation, clearly should be moved by water.

HYDROELECTRIC RESOURCES OF AMAZÔNIA

The lack of intermodal balance in Brazilian transportation, which is so strongly skewed in favor of highways, may be discerned as yet another instance of a recurrent sequence that involves the crowning of and eventual disenchantment with "king" solutions. It is disturbing, therefore, to find in the organ of the National Confederation of Commerce an article touching upon transportation in Amazônia that is entitled "Railroad, the new way" (Loiola, 1974)--no less disturbing for one who has viewed with misgiving the all-out highway program. The director of the National Department of Highways (DNER) under the new administration is reported by the author of the article to have announced that, although work on the Transamazon and the Perimetral highways is to proceed, it will be at a slower pace. But, where one might hope to discover a hint of a new equilibrium in the appraisal of transportation modes, the article predicts no less than the "death of the highway system by the close of the century." It goes on to forecast in vivid rhetoric that "railroads, which were proscribed from Amazônia, condemned and exorcised as enemies of the national economy," will make a comeback in that region, "where the greatest Brazilian hydraulic potential occurs. Recent studies carried out in Amazônia . . . showed amazing volumes of unused hydraulic power." One may hope that the "all-or-nothing," the "either-or" stance of the writer does not reflect official policy.

Whether or not future transportation in Amazônia is envisaged largely in terms of electrified trunk-line railroads--electric traction is generally considered most appropriate for dense traffic--there is no doubt that among the latest resources of the region to receive widespread recognition and whet the appetite of developers is hydroelectric power.

Immense possibilities are seen for harnessing major Amazonian tributaries, especially their upper and middle courses. Eletronorte, a subsidiary of Eletrobrás (the federal government power holding company), was established in 1973 to conduct research and development work in the region. What are no more than informed guesses, based on preliminary studies carried out up to mid-1974, suggest a total hydroelectric potential of some 60 million kW for the Brazilian portion of the Amazon basin proper and a further 18 million kW in the contiguous Araguaia-Tocantins basin. With little or no public understanding of or sympathy for a deliberate policy of slowing the growth rate of the utility industry, one may expect considerable pressure to develop these potentials as rapidly as possible.

Prospective uses of Amazônia's newly revealed hydro-power resources may include transportation (by rail) and transformation (in the region and its periphery) of iron mineral from the Serra dos Carajás, between the Xingu and Tocantins rivers (an estimated

18 billion tons of high-grade ore), as well as processing of cassiterite, especially in Rondônia (an estimated 5 million tons), and bauxite, which occurs north and south of the Amazon River. In addition to supplying an increasing demand within the region, hydro-power, produced at an ecological cost to Amazônia, is to benefit metropolitan areas in the northeast and southeast of Brazil, whereto it will be exported by means of interties with the electric grids of these regions. Most of the sites for hydroelectric plants in Amazônia are remote in respect to major existing markets, and therefore pose considerable technological problems for power conveyance over very long distances. New means of transmission of ultrahigh voltages are now being developed and doubtless these problems eventually will be solved.

Less certain is that the ecological cost of developing Amazonian hydroelectric resources will be adequately weighed--or even perceived in time. In the maturation of public concern with evironmental quality, the first stage appears to be one of disquiet about ill effects that can be experienced directly *e. g.*, chemical wastes and untreated sewage in streams, lakes and along the sea shore; smog in the air and the like. At this level of ecological awareness the key-word is "pollution" in the narrowest sense, and hydroelectric facilities--especially when contrasted with generating systems powered by traditional fuels--are perceived by most people as an environmentally-sound form of utilizing a renewable natural resource. At most, there may be some thought given to the large areas of land that are flooded in building the power plants, and to the immediate effects thereof. Long-range, often unforeseeable chain reactions produced by river-basin development projects rarely are considered. And yet, particularly in tropical environments, the price can be very high, as in the case of food sources eliminated or health hazards (*e. g.*, schistosomiasis) created or exacerbated.

In conclusion, the husbanding of energy, irrespective of whether derived from renewable or non-renewable resources, would seem to suggest that a considerable effort be made to promote the effective utilization of the Amazonian waterways within a judicious multi-modal framework. There is an infinite range in the degree of possible intervention in and resulting damage to the Amazon river ecosystem. At one extreme are certain proposals to tamper with the drainage that would undoubtedly have catastrophic effects, *e. g.*, damming up the Amazon to create the so-called "Amazon Sea." At the other extreme, there are a number of measures having very minor effects on the environment, such as installation of beacons and buoys, building of port facilities, removal of sunken logs, and adoption of up-to-date, special-type watercraft.

An awesome responsibility rests on the individuals whose value judgements and decisions--many irreversible--must reconcile the different interests that concern the Amazon system. Present scientific knowledge is insufficient, and yet alternatives must be weighed and combinations worked out that optimize the transport networks and minimize the hazards to the riverine environment. This has been generous to its original inhabitants, but appears highly vulnerable to the impact of modern temperate-zone technology.

ZUSAMMENFASSUNG

Der Überblick, der keinen Anspruch auf Vollständigkeit erhebt, behandelt das Amazonas-Becken im Bereich der im allgemeinen tiefer gelegenen äquatorialen und tropischen Gebiete Brasiliens.

Das erste Kapitel behandelt den Amazonas selbst. Der Strom wird in dieser Abhandlung zunächst vom Meer her betrachtet, wie dies die europäischen Entdecker, vermutlich im Jahre 1500, taten. Der Amazonas führt etwa 15 % des gesamten in die Ozeane fließenden Süßwassers, die mitgeführten Sedimente besitzen eine beträchtliche morphologische Bedeutung für die Küstenbildung der Guayanas.

Die Sedimentgesteine der Amazonasebene, die zwischen den alten Schilden Guayanas und Zentralbrasilien liegt, sind die obersten einer Schichtenfolge, die sich in einer Serie Ost-Westgerichteter Becken absetzte. Die langsame Senkung der Erdkruste entlang der Achse dieser Becken, die durch Hebung andernorts ausgeglichen wurde, führte zu einer Verbiegung und zu Brüchen des Sockels und der darüberliegenden Materialien. Nur im jüngsten Alluvium wird das hydrographische Netz nicht durch das daraus resultierende Mosaik von Klüften und Brüchen beherrscht.

In Amazonien gibt es zwei grundsätzlich verschiedene Gruppen von Oberflächenformen: die *terras firmes,* die nicht überschwemmten hochgelegenen Teile, und die *várzeas* oder Überschwemmungsebenen.

Es gibt zwar verschiedene Flächenniveaus im terra firme-Gebiet, aber über die Entstehung dieser Oberflächen herrscht nur wenig Übereinstimmung. Während des Pleistozän und möglicherweise noch im späten Tertiär war der Wechsel von Erosion und Aufschüttung an glazial-eustatische Transgressionen und Regressionen gebunden. Rezente alluviale Aufschüttungen füllten ertrunkene Täler auf und bildeten somit die heutige Stromaue. Trotzdem blieben große offene Wasserflächen im Bereich der Unterläufe verschiedener Nebenflüsse erhalten. Die Tatsache, daß einige größere Zuflüsse ihre seenartigen Mündungen teils aufschütten konnten, andere dagegen nicht, erklärt sich aus der Menge der mitgeführten Sedimente. Der Zusammenhang zwischen rein optischen Eigenschaften ("Weiß-, Klar- und Schwarzwasser"), dem relativen Reichtum an Sedimenten und den verschiedenen Quellgebieten ist erwiesen.

Die Überschwemmungsaue weist verschiedenartige Aspekte auf. Nebenflüsse im Alluvium bilden typische Mäanderformen. Der Amazonas selbst zeigt ein anderes Bild: Er gabelt sich auf, fließt wieder zusammen und umschließt dabei linsenförmige Inseln; das Pendeln der Haupt- und Seitenkanäle läßt langgestreckte Bündel von Anwachsstreifen und Rinnen entstehen, während in den tiefliegenden Teilen der Ebene die Überschwemmungen häufig zur Bildung von gewundenen Flußmustern führen.

Die alljährlichen Überschwemmungen, deren regelmäßiger Ablauf bemerkenswert ist, besitzen eine große ökologische und kulturelle Bedeutung. So ist das Wasser für die Verbreitung von várzea-Pflanzen verantwortlich, die auch eine entsprechende Anpassungsfähigkeit entwickelt haben. Seen und periodisch überflutetes Land dienen als Brutstätten für die Flußfauna und liefern gleichzeitig einen wertvollen Teil des Nahrungsaufkommens. Die säuerlichen Früchte der várzea bieten möglicherweise eine Vitamin C-Quelle für pflanzenfressende Fische. Da die Samen der Früchte unversehrt ausgeschieden werden, wird eine Verbreitung der Pflanzen bewirkt. Im Hinblick auf die besonderen und bisher unerkannten Wechselbeziehungen zwischen den einzelnen Komponenten des Ökosystems müssen Pläne, die drastische Veränderungen der jetzigen Gegebenheiten hervorrufen, mit großer Besorgnis betrachtet werden.

Wasserspiegelschwankungen sind von großer kulturgeographischer Bedeutung, besonders in der várzea. Extraktion der Flora und Fauna, Ackerbau und der Transport der landwirtschaftlichen Produkte zu den Märkten hängen von diesem Rhythmus ab.

Im zweiten Kapitel werden anhand typischer Beispiele die Nebenflüsse des Amazonas behandelt und ihre Bedeutung als Verkehrsadern sowohl vor als auch nach der Entdeckungszeit aufgezeigt. Während der Periode, in der die Monarchien der Iberischen Halbinsel vereinigt waren, sind die Portugiesen nach der Vertreibung englischer und holländischer Eindringlinge auf dem Amazonas und den größeren Nebenflüssen vorgestoßen und haben das spanische *dejure*-Territorium verkleinert. In der Mitte des 18. Jhs. sicherte eine Reihe von Befestigungsanlagen Portugals *de facto*-Besitzungen, wobei sich schon die Umrisse des heutigen brasilianischen Amazonien abzeichneten.

Ob ein bestimmter Nebenfluß zu einem Korridor für den Verkehr und die Besiedlung wurde oder ob er weiterhin den Indianern überlassen blieb, hing letzten Endes weniger von natürlichen Hindernissen für die Schiffahrt ab als von tatsächlichem oder vermutetem Vorhandensein von Bodenschätzen längs seines Laufs oder in seinem Quellgebiet.

Verschiedene Indianerstämme zogen sich in das durch zahlreiche Wasserfälle und Stromschnellen abgeschlossene Oberlaufgebiet des Xingú zurück, das von dürftiger Savannenvegetation umgeben war. Hier beeinflußten sich jahrhundertelang Vertreter der wichtigsten südamerikanischen Sprachgruppen und lieferten ein einzigartiges Beispiel der Akkulturation von verschiedenen Stämmen. In jüngster Zeit beeinträchtigten riesige Viehfazenden den Bereich des oberen Xingú-Beckens, und selbst die Unversehrbarkeit cines Indianerreservats wurde verletzt.

Die Diskussion über die Bevölkerungsdichte der Ureinwohner in präkolumbianischer Zeit in Amazonien und die spätere Entvölkerung konzentriert sich besonders auf das Tapajóz-Gebiet. *Terra preta,* ein schwarzer Boden, der zahlreiche Spuren ehemaliger Besiedlung aufweist, tritt häufig längs der Uferhänge auf. Schon zu Beginn des 19. Jhs. war jedoch die Tapajó-Nation "spurlos verschwunden" (Martius).

Der Rio Madeira eignete sich – wie der Tapajóz und der Xingú – als Zufahrtsweg zu den Goldlagerstätten im westlichen Zentralbrasilien und bot die zusätzliche Aussicht auf einen rückseitigen Zugang zu der sagenumwobenen Region Potosí. Es überrascht nicht, daß der Madeira ein Gegenstand politischer Manöver Portugals und Spaniens war. Nach der Unabhängigkeit der ehemaligen Kolonien wurden der Oberlauf des Madeira und die Quellflüsse der Nachbarströme nach Westen zu zum Schauplatz des Rohgummi-booms, der zu größeren Veränderungen in der Grenzziehung zwischen Bolivien und Brasilien führte.

Der Rio Negro nimmt eine Sonderstellung unter den Amazonas-Nebenflüssen ein. Sein optisches Erscheinungsbild, das durch wasserlösliche, von zersetztem Pflanzenmaterial stammende Farbstoffe verursacht wird, bildet seine hervorstechendste Eigenschaft. Haben Anwohner und Reisende den Mangel an Lebewesen in diesem "Hungerfluß" beklagt, so konnten sie andererseits auch eine relativ geringe Belästigung von gewissen Insektenplagen verzeichnen. Es wurde vermutet, daß der Rio Negro "voll von Insektiziden" sein muß, da er große Mengen von Pflanzensubstanzen besitzt, die als Schutz gegen pflanzenfressende Räuber dienen. Der Gedanke, daß der Rio Negro in sich selbst eine wertvolle Ressource darstellen könnte, ist völlig neu. Die Kolonialmächte sahen in ihm jahrhundertelang nur einen Durchgangsweg für ihre Eroberungspläne oder eine Hintertür, die verteidigt werden mußte.

Von den Wasserstraßen des Rio Negro-Systems hat keine eine größere Aufmerksamkeit verursacht als der Casiquiare-Kanal, der die Verbindung mit dem Orinoco herstellt. Teils wird angenommen, daß diese Verbindung das Ergebnis der rückschreitenden Erosion im Quellgebiet des Rio Negro-Systems sei, teils, daß sie vom Orinoco gefördert wurde, indem dieser Fluß seine Fluten über einen niedrigen Punkt der Wasserscheide ergoß. Krustenverschiebungen oder der Wechsel arider und humider Klimate können ebenfalls zu einer Verlagerung der Wasserscheide geführt werden.

Was auch immer der Ursprung des Casiquiare sein mag, er scheint eine der bedeutendsten präkolumbianischen Passagen Amerikas gewesen zu sein. Heute sind Pläne zur Verbesserung des natürlichen Kanals und zum Bau einer Alternativlösung vorhanden. Der jüngste Vorschlag für das erste Verbindungsstück in einem kontinentalen Schiffahrtssystem vom Orinoco zum La Plata wurde von der venezolanischen Regierung unterbreitet. Eine Studie zur Durchführbarkeit dieses Projekts schlägt sowohl die Verwendung von Atomkraft als auch die Anwendung konventioneller Mittel für die Bauarbeiten vor.

Der größte Nebenfluß des Rio Negro ist der Rio Branco, dessen Quellflüsse in einem Grasland- und Savannen-Gebiet zusammenlaufen.Im Osten wird das Tiefland zwischen dem Rio Branco und dem nordwärts fließenden Essequibo während der Regenzeit zu einem See; dieses Tiefland wird im allgemeinen als das Gebiet eines ehemals nach Osten orientierten Entwässerungssystems angesehen, das zum Amazonas abgeleitet wurde. Trotz des nur temporär auftretenden Sees wurden dessen Existenz zusammen mit der Legende von einer goldenen Stadt an den Ufern eines großen Binnensees zwischen Amazonas und Orinoco an der nördlichen Küste Südamerikas bekannt. Ein legendärer Schatz schürte die Habsucht der Regierungen und lockte zahlreiche Abenteurer aus Europa an. Die Portugiesen hielten ihr Interesse wach und versuchten, Forschungsexpeditionen zu stoppen. Ihre Strategie war durch die Tatsache beeinflußt, daß der Rio Negro nicht weit entfernt von der Mündung des Madeira in den Amazonas einmündet, der den Zugang zu den Bergbaugebieten von Mato Grosso und Cuiabá ermöglichte.

Es wurde für notwendig gehalten, den Oberlauf des Rio Branco zu besiedeln, und als Basis für diese Besiedlung wurde die Viehzucht vorgesehen. Die Weideflächen, die zur Viehzucht besonders geeignet schienen, haben jedoch den Erwartungen nicht entsprochen, und auch der Transport auf dem Fluß ist nur saisonal und mit Schwierigkeiten möglich. Eine Überlandverbindung (Manaus - Boa Vista) ist nun als Teil des panamerikanischen Straßensystems im Bau und wird von der seit 1973 im Bau befindlichen Perimetral do Norte gekreuzt.

Das dritte Kapitel ist dem Transportwesen im brasilianischen Amazonien gewidmet, unter besonderer Berücksichtigung des gigantischen Fernstraßennetzes, das jetzt über das amazonische Flußsystem mit seinen 20.000 fahrbaren Flußkilometern gelegt wird. Während die Schiffbarkeitsmöglichkeiten in diesem Flußnetz nicht so groß sind, wie beim Blick auf die Karte vermutet werden könnte, sind die sehr beträchtlichen geographischen Vorzüge nicht ausgewertet worden. Im Hinblick auf die hohen Kosten ineffizienten Betriebs sind die Spediteure dabei, die Wasserwege zugunsten des schnelleren und insgesamt gesehen billigeren LKW-Transports aufzugeben, sobald die Alternative gegeben ist. Vor 20 Jahren glaubte man nur an die Möglichkeit des Transports auf dem Wasserwege, um die ungeheuren Entfernungen Amazoniens zu bezwingen, ohne die Frachtkosten übermäßig zu erhöhen. In jüngster Zeit kommt den Wasserwegen nur noch eine Ergänzungsfunktion zu, und aktuelle Pläne sehen ein Straßennetz von 17.855 km vor. Die Vorteile, die der Wassertransport für gewisse Frachtarten bietet, wurden durch die Ungeduld der Entwicklungsplaner aufgehoben. Obwohl die wirtschaftliche Effektivität des amazonischen Straßennetzes nach dem Wert der Bodenschätze beurteilt werden wird, deren Entdeckung und Ausbeutung in den bisher unzugänglichen Gebieten zwischen den Nebenflüssen nun möglich wird, es ist unrealistisch, die Straßenbauprogramme dieser Region auf rein ökonomischer, geschweige denn ökologischer Ebene zu sehen. Die Länder, die am Amazonasbecken einen Anteil besitzen, haben bislang ihren Hylaea-Anteil noch nicht effektiv in ihren Staatsraum integriert. Sie bemühen sich um die Erschließung dieser abgelegenen Teile ihres Territoriums, und viele offizielle Verlautbarungen setzen die Kolonisation Amazoniens mit nationaler Sicherheit gleich. Jedoch sind nicht-militärische Faktoren mit der nationalen Politik verflochten. So gab die schlechte Situation der Bevölkerung des nordostbrasilianischen Hinterlandes, die 1970 durch eine weitere Trockenheit betroffen wurde, den unmittelbaren Anstoß zur Aufstellung des brasilianischen Programms der "nationalen Integration", dessen Eckpfeiler die Transamazônica und die Cuiabá-Santarém-Straße sind. Mit dem Schwerpunkt auf der Besiedlung Amazoniens wurden umfangreiche Gebiete entlang der Fernstraßen für eine Kolonisation und Agrarreform reserviert. Etwa 2 1/4 Millionen qkm stehen zur Kolonisation zur Verfügung, ein Gebiet, das nur von 10 Ländern dieser Erde flächenmäßig übertroffen wird. Während die ursprüngliche Entwicklungspolitik den Familienbetrieb begünstigte, scheint die neue Regierung, die ihr Amt 1974 antrat, den Schwerpunkt auf die wirtschaftliche Erschließung Amazoniens durch große Konzerne zu legen.

Trotz des deklarierten Ziels einer "perfekten Straßen-Fluß-Verbindung" hat das beschränkte Transportprogramm für Amazonien für den Bau von Flußhäfen, die Verbesserung der Wasserwege und die Anschaffung geeigneter Wasserfahrzeuge verhältnismäßig geringere Mittel bereitgestellt als für den Bau von Straßen und Flugplätzen. Eine solche Politik steht im Gegensatz zu der Notwendigkeit des Energie-Sparens und erhöht die Abhängigkeit des Landes von ausländischen Ölquellen. Die beschränkte Förderung von Transportarten ist ein weiteres Beispiel von Begeisterung und Ernüchterung im Hinblick auf ein jeweiliges Allheilmittel. Ein Autor, der schon vom "Tod" des Fernstraßennetzes gegen Ende des Jahrhunderts spricht, sagt der Entwicklung elektrifizierter Eisenbahnlinien eine brillante Zukunft voraus, die auf der Nutzung des ungeheuren Wasserkraftpotentials basiert, das in Amazonien entdeckt wird. In der Tat ist eine der jüngst anerkannten und von den Entwicklungsplanern weithin hoch eingeschätzten Ressourcen der Region die Wasserkraft. Es erhebt sich die Frage, ob die ökologischen Nachteile der Entwicklung amazonischer Wasserkraftreserven rechtzeitig entsprechend geprüft oder überhaupt erkannt werden.

Eine schwere Verantwortung ruht auf den Menschen, deren Bewertung und Entscheidungen – von denen viele unwiderruflich sind – die verschiedenen Interessen, die das Amazonas-System betreffen, in Einklang bringen müssen. Die Flußlandschaft war den Ureinwohnern gegenüber großzügig, scheint aber auf den Eingriff der modernen Technologie der gemäßigten Zone äußerst verwundbar zu reagieren.

Übersetzung: G. Kohlhepp

REFERENCES

Ab'Saber, A. N. (1967) Problemas Geomorfológicos da Amazônia Brasileira. *In* Lent, H. (ed.) *Atas do Simpósio sôbre a Biota Amazônica. I-Geociências.* Pp 35–67. Conselho Nacional de Pesquisas, Rio de Janeiro.

A Crise Energética Mundial (1973) *Conjuntura Econômica* 27 (6): 105–117.

Acuña, C. de (1641) *Nuevo descubrimiento del gran Rio de las Amazonas . . . 1639.* Imprenta del Reyno, Madrid. Reproduced in C. M. de Almeida (1874) *Memorias para a Historia do Extincto Estado do Maranhão* . . . Hildebrandt, Rio de Janeiro.

Almada, M. da G. L. d' (1786) Letter to João Pereira Caldas. Barcellos, October 31, 1786. *Cit. in* Reis, A. C. F. (1940).

Almeida, L. A. (1967) Barragem da Paz–Origem da ideia. *Bol. geogr.* Rio de J. 26 (201): 35–40.

Almeida, W. F. (1974) Aspectos Toxicológicos dos herbicidas 2, 4-D e 2, 4, 5-T. *Biológico.* 40 (2): 44–51.

Azevedo, P. (1970) A Transamazônica como Fator de Desenvolvimento da Pesca Amazônica. *Equipesca J.* 7(35): 2–4.

Barat, J. (1973) Política de Transportes: avaliação e perspectivas face ao atual estágio de desenvolvimento do País. *Rev. bras. Econ.* 27(4): 51–83.

Barata F. (1950) *A Arte Oleira dos Tapajó. Considerações sôbre a cerâmica e dois tipos de vasos característicos.* Instituto de Antropologia e Etnologia do Pará (Publ. nº 2) Belém.

Barbosa, O. and J. R. A. Ramos (1959) *Território do Rio Branco. Bol. Dep. nac. Prod. min.* nº 196, Rio de J.

Bischoff, G. (1963a) Mitteilung zur Geologie des Territoriums Acre im Grenzgebiet von Brasilien und Peru. *Geol. Jb.* 80: 795–800.

-- (1963b) Zur Geologie des Amazonasbeckens. *Geol. Jb.* 80: 771–794.

Borah, W. (1970) The Historical Demography of Latin America: Sources, Techniques, Controversies, Yields. *In* De Prez, P. (ed.) *Population and Economics* (Proc. Sect. V, 4th Congr. Int. Econ. Hist. Assoc. 1968). University of Manitoba Press, Winnipeg. Pp. 173–205.

Bowers, W. S. (1971) Juvenile Hormones. *In* Jacobson, M. and D. G. Crosby (eds.) *Naturally occurring insecticides.* Marcel Dekker, Inc., New York. Pp. 307–332.

Brasil (1949) *Anais da Comissão Especial do Plano de Valorização Econômica da Amazônia.* (vol. 3) Câmara dos Deputados. Rio de Janeiro.

-- (1968a) *Hidrovias e Interligação de Bacias Hidrográficas.* Ministério das Relações Exteriores. Rio de Janeiro.

-- (1968b) *Medição de Descarga e seus Problemas Técnicos no Maior Rio do Mundo.* Ministério das Minas e Energia, Departmento Nacional de Águas e Energia. Rio de Janeiro.

-- (1973) Embaixada do Brasil. *Bol. Esp.* nº 28. 26 de julho de 1973. Washington.

Bremer, H. (1973) Der Formungsmechanismus im tropischen Regenwald Amazoniens. *Z. Geomorph. N. F.* 17: 195–222.

Broecker, W. S., J. C. Kulp and C. S. Tucek (1956) Lamont Natural Radiocarbon Measurements III. *Sci.* 124(3213): 154–165.

Brooks, E., R. Fuerst, J. Hemming and F. Huxley (1973). *Tribes of the Amazon Basin in Brazil 1972. Report for the Aborigines Protection Society.* Charles Knight & Co. Ltd. London and Tonbridge.

Brown, C. B. and W. Lidstone (1878) *Fifteen Thousand Miles on the Amazon and its Tributaries.* Edward Stanford, London.

Burnier, L. P. (1971) *O Lago da Paz na Integração Geopolítica e Econômica das Regiões Centro-Oeste e Norte do Brasil.* Reprinted from *Mar* (Bol. Clube Naval) 211, Rio de. J.

Camargo, F. C. (1958) Report on the Amazon Region *In* UNESCO, *Problems of Humid Tropical Regions/Problèmes des régions tropicales humides.* Paris.

Carta Regia . . . (1753) Carta Regia to Governor F. X. de Mendonça Furtado. Lisboa April 25, 1753. *In* Mendonça, M. C. de (ed.) (1963) *A Amazônia na Era Pombalina.* (Vol.I) Instituto Histórico e Geográfico Brasileiro. Rio de Janeiro.

Casement, R. (1911–12) Correspondence respecting the Treatment of British Colonial Subjects and Native Indians Employed in the Collection of Rubber in the Putumayo District. *In* Great Britain *Parliamentary Publications (1912–1913)* Vol. 68. Cd. 6266. H. M. S. O.

Chatterjee, I. B. (1973). Evolution and the Biosynthesis of Ascorbic Acid. *Sci.* 182 (4118): 1271–1272.

Cooke, H. B. S. (1972) Pleistocene Chronology: Long or Short? *Marit. Sediments* 8 (1): 1–12.

Correa, F. P. (1777) *Anal de Villa Bella do ano de 1777. Cit. in* Freyre, G. (1968) *Contribuição para uma Sociologia da Biografia. O exemplo de Luiz de Albuquerque, Governador de Mato Grosso no fim do século XVIII. I. Comentário.* Academia Internacional da Cultura Portuguesa. Lisboa.

Corrêa Filho, V. (1969) *História de Mato Grosso.* Instituto Nacional do Livro, Ministério da Educação e Cultura, Rio de Janeiro.

Crevaux, J. (1883). *Voyages dans l'Amerique du Sud.* Librairie Hachette et Cie., Paris.

Cunha, E. da (1906) *Relatorio da Commissão Mixta Brasileiro-Peruana de Reconhecimento do Alto Purus. Notas Complementares do Commissario Brasileiro. 1904–1905.* Imprensa Nacional, Rio de Janeiro. Republished (1960) *O Rio Purus,* SPVEA (Superintendência do Desenvolvimento da Amazônia), Rio de Janeiro.

-- (1907) *Contrastes e Confrontos.* Emp. Litteraria e Typographica, Porto.

-- (1909) *A Marjem da Historia.* Livraria Chardron de Lello & Irmão, Porto.

Curiel R., J. (1971) *El desarrollo de la Región Sur de Venezuela – su importancia para la intercomunicación de los grandes rios de Suramérica y como vehículo de cooperación y desarrollo del sur del continente.* MOP. Secretaría General, Oficina de Información y Relaciones Publicas, Caracas.

Dobyns, H. F. (1966) Estimating Aboriginal American Population. 1. An Appraisal of Techniques with a New Hemispheric Estimate. *Curr. Anthrop.* 7(4): 395–416.

Ducke, A. (1949) Árvores Amazônicas e sua Propagacao . . . *Bol. Mus. para. 'E. Goeldi' Hist. nat. Ethnogr.* 10:81–92.

Dudley, Sir Robert (1646–47) *Dell'arcano del Mare. F. Onofri, Firenze.* (3 Vols.) The map is found in v. 2. According to P. L. Phillips, *A list of Geographical Atlases in the Library of Congress* (1909), three manuscript volumes, the first two dated 1610, formerly preserved in Florence, may be the basis of this work.

Ecotécnica (ca. 1967) *Indústria e Pasteurização de Leite do Amazonas Limitada.* (typescript). IPLAM, Manaus.

Eden, M. J. (1971) Scientific Exploration in Venezuelan Amazonas. *Geogl. J.* 137(2): 149–156.

Edmundson, G. (1903) The Dutch on the Amazon and Negro in the Seventeenth Century.

Engl. Hist. Rev. 18: 642–663.

El Ejercito lleva la revolución a la selva. *Oiga* 13 (568): 14–15, 38, 40.

El Peruano (Diario Oficial) 15 de Junio 1965. Decreto Supremo: Se aprueba el Programa de Núcleos Selváticos en las zonas fronterizas del país.

Falesi, I. C. (1967) O Estado atual dos Conhecimentos sôbre os Solos da Amazônia Brasileira. *In* Lent, H. (ed.) *Atas do Simpósio sôbre a Biota Amazônica. I-Geociências.* Pp. 151–168. Conselho Nacional de Pesquisas, Rio de Janeiro.

Ferreira, A. R. (1786) Informação . . . sobre o cumprimento que deu á ordem recebida de João Pereira Caldas para fazer reconhecimentos nas Povoações da parte inferior do Rio Negro e nas de novo estabelecidas no Rio Branco – 10 de Agosto 1786. *In* Brazil (1903) *Questions de limites soumises a l'arbitrage de S. M. le Roi d'Italie par le Brésil et la Grande-Bretagne.* Annexes du Premier Mémoire du Brésil, Vol. I. Documents d'origine portugaise. Doc. No. 75. Pp. 215–218.

Ferreira, E. O., F. F. M. de Almeida, E. F. Suszczynski and G. R. Derze (1971) *Mapa Tectônico do Brasil* (1:5,000,000). Departamento Nacional da Produção Mineral, Rio de Janeiro.

Fonseca, J. S. and P. A. Almeida (1899) *Voyage autour du Brésil.* (Édition pour les Americanistes) Librairie A. Lavignasse Filho & Cia., Rio de Janeiro.

Galvão, E. and M. F. Simões (1964) Kulturwandel und Stammesüberleben am oberen Xingu, Zentralbrasilien. *Beiträge zur Völkerkunde Südamerikas;* Festgabe für Herbert Baldus zum 65. Geburtstag. Völkerkundliche Abhandlungen: Band I. Becher, H. (ed.) Hannover. Pp. 131–151. Original version in Portuguese, Mudança e sobrevivência no Alto Xingu, Brasil Central, published in 1966 *Rev. Antrop.* 14: 37–52 and reproduced in E. Schaden (ed.) (1972) *Homem, Cultura e Sociedade no Brasil.* Editora Vozes Ltd. Petrópolis.

Garner, H. F. (1966) Dérangement of the Rio Caroni, Venezuela. *Revue Géomorph. dyn.* 16(2): 54–83.

-- (1967) Rivers in the Making. *Scient. Am.* 216(4): 84–94.

Gessner, F. (1959) *Hydrobotanik–Die Physiologischen Grundlagen der Pflanzenverbreitung im Wasser.* (Vol. 2) VEB Deutscher Verlag der Wissenschaften, Berlin.

-- (1962) Der Elektrolytgehalt des Amazonas. *Arch. Hydrobiol.* 58(4): 490–499.

Gibbs, R. J. (1965) *The Geochemistry of the Amazon River (Basin).* Dissertation, University of California, San Diego.

-- (1967) The Geochemistry of the Amazon River System: Part I. The factors that control the salinity and the composition and concentration of the suspended solids. *Bull. geol. Soc. Am.* 78: 1203–1232.

Hartt, C. F. (1874). Report of a reconnaissance of the Lower Tapajos [Morgan Expeditions 1870–1871]. *Bull. Cornell Univ. (Science)* 1(1): 11–37.

-- (1885) Contribuições para a Ethnologia do Valle do Amazonas. *Arch. Mus. nac. Rio de J.* 6:1–174

-- (1896) A Geologia do Pará. *Bol. Mus. para. 'E. Goeldi' Hist. nat. Ethnogr.* 1(3): 257–273.

Heriarte, M. (1964 [ca. 1662]) *Descriçam do Estado do Maranham, Para, Corupa, Rio das Amazonas.* Faksimile-Ausgabe aus den MSS 5880 und 5879 der Österreichischen National-Bibliothek, Wien. Einleitung Karl Anton Nowotny-Wien. Akademische Druck- und Verlagsanstalt, Graz.

Hirst, E. (1973a) Energy-Intensiveness of Transportation. *Transport. Eng. J.* (Proc. Ame. Soc. Civ. Eng.) 99 (TEl): 111–122.

-- (1973b) *Transportation Energy Conservation: Opportunities and Policy Issues.* (Testimony submitted to the U. S. House of Representatives Subcommittee on Conservation and Natural Resources [Committee on Government Operations] and Subcommittee on Energy [Committee on Science and Astronautics] pursuant to hearings on Conservation and Efficient Use of Energy, July 1973.)

Huber, J. (1909). Mattas e Madeiras Amazonicas. *Bol. Mus. para. 'E. Goeldi' Hist. Nat. Ethnogr.* 6: 91–225

Humboldt, A. de (1820–1822) *Voyage aux Regions Equinoxiales du Nouveau Continent fait en 1799, 1800, 1801, 1802, 1803 et 1804, par Alexander de Humboldt et A. Bonpland.* Chez N. Maze, Libraire, Paris. (Vols. 6 and 7, 1820; Vol. 8, 1822).

Hydraulics Laboratory (1962) *Demerara Coastal Investigation; Report on Siltation of Demerara Bar Channel and Coastal Erosion in British Guiana.* Delft.

INCRA (Instituto Nacional de Colonização e Reforma Agrária) (1971) Novas aspirações para um antigo caminho. Diretrizes básicas para ação do Ministério da Agricultura, através do INCRA na Transamazônica. (Estudos preliminares).

-- (1972) *Altamira 1.* Rio de Janeiro.

-- (ND) *Areas da Região Amazônica a serem discriminadas (Dec. Lei Nº 1164 de 1–4–1971).* Map at 1:5,000,000. INCRA–DF, Divisao de Cartografia. Not publ.

Junk, W. (1970) Investigations on the Ecology and Production-Biology of the "Floating Meadows" *(Paspalo-Echinochloetum)* on the Middle Amazon. *Amazoniana* 2(4): 449–495.

Junqueira, C. (1973) *The Brazilian Indigenous Problem and Policy: the Example of the Xingu National Park.* International Work Group for Indigenous Affairs, Doc. 13. Copenhagen/Geneva.

Kafatos, F. C., C. M. Williams and D. Prescott (1967) Insect hormone analogues in the black waters of the Rio Negro. *R/V 'Alpha Helix' Amazon Expedition February to October, 1967.* U. S. National Science Foundation/Scripps Institution of Oceanography, University of California, San Diego.

Kanwisher, J. and D. Prescott (1967) Analysis of Rio Negro Water. *R/V 'Alpha Helix'* Amazon Expedition February to October, 1967. U. S. National Science Foundation/Scripps Institution of Oceanography, University of California, San Diego.

Keller, F. (1875) *The Amazon and Madeira Rivers. Sketches and Descriptions from the Note-Book of an Explorer.* J. B. Lippincott and Co., Philadelphia.

Kemys, L. (1596) *A Relation of the Second Voyage to Guiana.* Thomas Dawson, London.

Klammer, G. (1971) Über plio-pleistozäne Terrassen und ihre Sedimente im unteren Amazonasgebiet. *Z. Geomorph. N. F.* 15(1): 62–106.

Koch-Grünberg, T. (ND) *cit. in* E. Nordenskiöld (1916) *op. cit.*

La Condamine, M. de (1749) Relation Abrégée D'un Voyage fait dans l'interieur de l'Amerique méridionale, depuis la Côte de la Mer du Sud, jusques aux Côtes du Brésil & de la Guiana, en descendant la rivière des Amazones. *Histoire de l'Académie Royale des Sciences, Année 1745.* Imprimerie Royale, Paris. pp. 391–492.

Landsberg, H. H. (1974) Low-Cost, Abundant Energy: Paradise Lost? *Sci.* 184(4134): 247–253.

Lathrap, D. W. (1970) *The Upper Amazon.* Praeger Publishers, New York.

Lochead, W. (1798 [1794]) Observations on the Natural History of Guiana. *Trans. R. Soc. Edinb.* 4(2): 41–63.

Loiola, G. (1974) Ferrovia, o novo caminho. *Comércio & Mercados* 8(80): 2–4.
López, V. M. (1956) Venezuelan Guiana. *In* Jenks, F. W. (ed.) *Handbook of South American Geology -- An Explanation of the Geologic Map of South America.* Mem. geol. Soc. Am. 65: 331–340.
Lowe-McConnell, R. H. (1967) Factors Affecting Fish Populations in Amazonian Waters. *In* Lent, H. (ed.) *Atas do Simpósio sôbre a Biota Amazônica. 7-Conservação da Natureza e Recursos Naturais.* Pp. 177–186. Conselho Nacional de Pesquisas, Rio de Janeiro.
Lyell, C. (1832) *Principles of Geology.* (Vol. I) (2nd ed.) John Murray, London.
Mabessone, J. M. (1967) Sedimentos correlativos do clima tropical. *In* Lent, H. (ed.) *Atas do Simpósio sôbre a Biota Amazônica. I – Geociências.* Pp. 327–337. Conselho Nacional de Pesquisas, Rio de Janeiro.
Marlier, G. (1967) Hydrobiology in the Amazon Region. *In* Lent, H. (ed.) *Atas do Simpósio sôbre a Biota Amazônica. III – Limnologia.* Pp. 1–7. Conselho Nacional de Pesquisas, Rio de Janeiro.
-- (1973) Limnology of the Congo and Amazon Rivers. *In* Meggers, B. J., E. S. Ayensu and W. D. Duckworth (eds.) *Tropical Forest Ecosystems in Africa and South America: A Comparative Review.* Smithsonian Institution Press, Washington D. C. Pp. 223–238.
Martius, C. F. P. v. (1867) Beiträge zur Ethnographie und Sprachenkunde Amerika zumal Brasiliens. I – Zur Ethnographie. Friedrich Fleischer, Leipzig.
Matsui, E., E. Salati, W. L. F. Brinkmann and I. Friedman (1972) [1973] Vazões relativas dos rios Negro e Solimões através das concentrações de ^{18}O. *Acta Amazonica* 2(3): 31–46.
Mattos, F. J. de (1949) Ante-Projeto para o Plano Nacional de Viação Fluvial (Linhas Mestras). Estabelecimento Gráfico Villani & Filhos Ltds. Rio de Janeiro. Reprinted from *Rev. Clube Eng.* (151): 49–66.
McConnell, R. B., D. Masson Smith and J. P. Berrange (1969) Geological and Geophysical evidence for a rift valley in the Guiana Shield. *Geol. Mijnbouw 48:* 189–199.
Médici, E. G. (1970) Statement by the President of Brazil quoted in Andreazza, M., Transamazônica – Pronunciamento feito na Câmara dos Deputados em 1º de julho de 1970. DNER (Dept. Nac. de Estrada de Rodagem) Rio de Janeiro.
Meggers, B. J. (1954) Environmental Limitations on the Development of Culture. *Ame. Anthropol.* 56(5): 801–824.
Memorial . . . (1974) Memorial presented to Governor Fragelli of Mato Grosso State. dated May 16, 1972 and signed by representatives of the Masonic Lodge and ten additional fraternal, service, business and amateur fish-and-game organizations of the Cuiabá community. (Typescript).
Nagell, R. H. (1961) The Serra do Navio Manganese district: a residual lateritic deposit. *Econ. Geol.* 56(7): 1333–1334.
Nimuendajú, C. (1949 [1939]) Os Tapajó. *Bol. Mus. para. 'E. Goeldi' Hist. nat. Ethnogr.* 10: 93–106.
Nordenskiöld, E. (1916) Die Anpassung der Indianer an die Verhältnisse in den Überschwemmungsgebieten in Südamerika. *Ymer* 36(2): 138–155.
Oltman R. E. (1968) *Reconnaissance Investigations of the Discharge and Water Quality of the Amazon River.* U. S. G. S. Circular 552, Washington.
-- (1973) *Personal Communication.* March 12, 1973.
Oltman, R. E., H. O'R. Sternberg, F. C. Ames and L. C. Davis, Jr. (1964) *Amazon River*

Investigations, Reconnaissance Measurements of July 1963. U. S. G. S. Circular 486. Washington.

Oyenhausen, J. C. A. (1811) Ofício nº17, Dirigido ao Conde Linhares . . . relativamente aos meios de comunicação da Capitania de Mato Grosso com as outras por via fluvial. *In* Hollanda, S. B. (1945) *Monções,* Livraria Editora da Casa do Estudante do Brasil, Rio de Janeiro.

Paiva, G. (1929) *Valle do Rio Negro (Physiographia e Geologia). Bol. Serv. geol. miner. Bras.* nº 40, Rio de Janeiro.

Palmatary, H. C. (1960) The Archaelogy of the Lower Tapajos Valley, Brazil. *Trans. Am. Philos. Soc.* N. S. 50 (3): 1–243.

Panero, R. (1967a) *On the use of low dams as a possible stimulant to South American development.* HI–788–RR. Hudson Institute, Croton-on-Hudson, New York.

-- (1967b) *A South American "Great Lakes" System.* HI–788/3–RR. Hudson Institute, Croton-on-Hudson, New York.

Passarge, S. (1931) Das Rio Branco-Essequibo-Problem. *Petermanns Mitteilungen* 77(5/6): 135–137.

Paz-Castillo, A. and P. Kruger (1972) *Design of a waterway connecting the Orinoco and Rio Negro Rivers in the Federal Territory of Amazonas, Venezuela.* Techn. Report nº. 153. Dept. of Civil Eng., Stanford University, Stanford, California.

Pereira, A. (1889) Relação do que ha no grande rio das Amazonas novamente descuberto. Año de 1616. *In* M. J. de la Espada (1889) *Viaje del Capitán Pedro Texeira Aguas Arriba del Rio de las Amazonas (1638–1639).* Imprenta de Fortanet, Madrid. Pp. 115–119.

Pombal, S. J. de C. Mello, Marquess of (1755) Letter of the Secretary of State for Foreign Affairs and War to his brother Governor F. X. de Mendonça Furtado March 17, 1755. *In* Mendonça, M. C. de (1963) *A Amazônia na Era Pombalina* (Vol. II.) Instituto Histórico e Geográfico Brasileiro. Rio de Janeiro.

Ralegh, W. (1596) *The Discoverie of the Large, Rich and Beuutiful Empyre of Guiana . . . Performed in the yeare 1595.* Robinson, London.

Recursos energéticos--aceleram-se as pesquisas (1974) *Conjuntura Econômica* 28(1): 72–78.

Reis, A. C. F. (1937) *A questão do Acre.* Typographia Phoenix, Manaus.

-- (1940) *Lobo d'Almada: Um Estadista Colonial.* 2nd ed. Imprensa Pública, Manaus.

-- (1947) *Limites e Demarcações na Amazônia Brasileira.* 1º (Tomo) Public. Comissão Brasileira Demarcadora de Limites. Imprensa Nacional, Rio de Janeiro.

-- (1960) *A Amazônia e a Cobiça Internacional.* Companhia Editôra Nacional. S. Paulo.

Reyne, A. (1961) On the Contribution of the Amazon River to Accretion of the Coast of the Guianas. *Geol. Mijnbouw* 40: 219–226.

Rice, H. (1921) The Rio Negro, the Casiquiare Canal, and the Upper Orinoco, September 1919– April 1920. *The Geogrl. J.* 58 (5): 321–359.

Ryther, J. H., D. W. Menzel and N. Corwin (1967) Influence of the Amazon River Outflow on the Ecology of the Western Tropical Atlantic. I. Hydrography and Nutrient Chemistry. *J. Mar. Res.* 25(1): 69–83.

Sakamoto, T. (1957) *Trabalhos Sedimentológicos, Geomorfológicos e Pedogenéticos referentes à Amazônia.* Missão FAO/UNESCO na Amazônia. SPVEA, Belém.

-- (1960) Rock Weathering on "Terras Firmes" and Deposition on "Varzeas" in the Amazon. *J. Fac. Sci.*, Univ. Tokyo. Section II. 12(2): 155–216.

Sampaio, F. X. R. de (1825) *Diario da Viagem . . . 1774–1775.* Tipografia da Academia, Lisboa.

Schaden, E. (1969) *Aculturação Indígena: Ensaio sôbre Fatôres e Tendências da Mudança Cultural de Tribos Índias em contacto com o Mundo dos Brancos.* Livraria Pioneira Editôra/ Editôra da Universidade de São Paulo, S. Paulo.

Scudder, T. *et al.* (1972) II – Irrigation and Water Development. Thayer Scudder, Chairman. *In* Farvar, M. T. and J. P. Milton (eds.) *The Careless Technology; Ecology and International Development.* Record of Conference on the Ecological Aspects of International Development. The Natural History Press, Garden City, N. J. Pp. 155–367.

Serra, R. F. de A. (1858) Memoria, ou Informação dada ao Governo sobre a Capitania de Mato Grosso . . . em 31 de Janeiro de 1800. *Rev. trimens. Hist. Geogr.* (Tomo Segundo) (2nd ed.). Rio de Janeiro.

SGTE-LASA (1971) Ligação Paraguai - Guaporé. *In* Dept. Nac. Portos e Vias Navegáveis, *Estudo Geral das Vias Navegáveis Interiores do Brasil – Vol. 11: Ligações de Bacias.* Rio de Janeiro.

Shipboard Scientific Party (1971) Leg 4 of the Deep Sea Drilling Project. *Sci.* 172(3989): 1197–1205.

Sinha, N. K. P. (1968) *Geomorphic evolution of the northern Rupununi Basin, Guyana.* McGill University Savanna Research Project (Savanna Research Series No. 11) Montreal.

Sioli, H. (1953a) Limnologische Untersuchungen und Betrachtungen zur erstmaligen Entdeckung endemischer Schistosomiasis (Sch. mansoni) im Amazonasgebiet. *Arch. Hydrobiol.* 48(1): 1–23.

-- (1953b) Schistosomiasis and Limnology in the Amazon Region. *Am. J. Trop. Med. and Hyg.* 2(4): 700–707.

-- (1954) Gewässerchemie und Vorgänge in den Böden im Amazonasgebiet. *Naturwissenschaften* 41(19): 456–457.

Sioli, H., G. H. Schwabe and H. Klinge (1969) Limnological Outlooks on Landscape-Ecology in Latin America. *Trop. Ecol.* 10 (1): 72–82.

Smith, H. H. (1879) *Brazil; The Amazons and the Coast.* Charles Scribner's Sons. New York.

Smoot, G. F. (1972) *Measurement of Discharge by the Moving-boat Method – June 5, 1972 – Amazon River at Óbidos, Brazil.* (Unpubl. typescript).

SNAPP (Serviço de Navegação do Amazonas e de Administração do Porto do Pará) (1951) *Subsídio para o Plano da Valorização Econômica do Vale Amazônico.* Departamento de Imprensa Nacional, Rio de Janeiro.

Sombroek, W. G. (1966) *Amazon Soils: A Reconnaissance of the Soils of the Brazilian Amazon Region.* Centre for Agricultural Publications and Documentation. Wageningen.

Souza, B. L. de (1959) *Do Rio Negro ao Orenoco (A Terra – O Homem).* Conselho Nacional de Proteção aos Índios, Rio de Janeiro.

Spix, J. B. von and C. F. P. von Martius (1823–1831) *Reise in Brasilien auf Befehl Sr. Majestät Maximilian Joseph I. Königs von Baiern in den Jahren 1817 bis 1820 gemacht und beschrieben* (3 vols.) M. Lindauer, München.

Spruce, R. (1908) *Notes of a Botanist on the Amazon and Andes . . . during the years 1849–1864.* (Ed. and condensed by A. R. Wallace). 2 vols. MacMillan and Company Ltd., London.

SPVEA (Superintendência do Plano de Valorização Econômica da Amazônia) (1955) *Primeiro Plano Quinquenal.* 2nd vol. Setor de Coordenação e Divulgação, SPVEA, Rio de Janeiro.

Stephanes, R. (1972) O Programa de Integração Nacional e a Colonização da Amazônia. (2nd ed.). INCRA, Brasília.

Stern, K. M. (1970) Der Casiquiare-Kanal, einst und jetzt. *Amazoniana* 2(4): 401–416.

Sternberg, H. O'R. (1955) Sismicité et morphologie en Amazonie Brésilienne. *Ann. Géogr.* 64(342): 97–105.

-- (1956) *A Água e o Homem na Várzea do Careiro.* (2 vols.) Rio de Janeiro.

-- (1960) Radiocarbon dating as applied to a problem of Amazonian morphology. *Comptes Rendus 18ème Congrès International de Géographie,* International Geographical Union, Rio de Janeiro.

-- (1965) Brazil: Complex Giant. *Foreign Affairs* 43(2): 297–311.

-- (1966) Die Viehzucht im Careiro-Cambixegebiet -- ein Beitrag zur Kulturgeographie der Amazonasniederung, Heidelberger Studien zur Kulturgeographie, Festgabe für Gottfried Pfeifer, Franz Steiner Verlag GmbH, Wiesbaden. Pp. 171–197.

-- (1973) Development and Conservation. *Erdk.* 27(4): 253–265.

Sternberg H. O'R. and R. J. Russell (1957) Fracture Patterns in the Amazon and Mississippi Valleys. *Proceedings 8th General Assembly and 17th International Congress.* International Geographical Union, Washington, D. C. Pp. 380–385.

Stokes, C. J. (1966) The Economic Impact of the Carretera Marginal de la Selva. *Traffic Q.* 20(2): 203–226. Reprod. in Stokes, C. J. (1968) *Transportation and Economic Development in Latin America.* Frederick A. Praeger, New York.

SUDAM (Superintendência do Desenvolvimento da Amazônia) (1967) *Primeiro Plano Regional de Desenvolvimento (1967–1971).* Serviço de Documentação e Divulgação, SUDAM, Belém.

-- (1971a) *Subsídios ao Plano Regional de Desenvolvimento (1972–1974).* SUDAM, Belém.

-- (1971b) *Plano de Desenvolvimento da Amazônia (1972–1974).* SUDAM, Belém.

Tastevin, C. (1925) Le fleuve Murú. Ses habitants. Croyances et moeurs *kachinaua. Géographie* 43(4–5): 403–422.

-- (1926) Le Haut Tarauacá. *Géographie* 45(1–2): 34–54 and 45 (3–4): 158–175.

TRANSCON/Berger (Consultoria Brasileira de Transportes Ltd. and Louis Berger Eng. Ltd.) (1968) *Engineering and Economic Feasibility Study BR–319. Highway Porto Velho-Manaus.* (Vol. 1) Departamento de Estradas de Rodagem do Amazonas, Manaus.

U.S. Army Corps of Engineers (1943) *Report on Orinoco-Casiquiare-Negro Waterway, Venezuela-Colombia-Brazil.* (4 vols.) Prepared for the Coordinator of Inter-American Affairs, Washington.

U.S.G.S. (United States Geological Survey) (1972) World Record River Flow Measured on Amazon. *Department of the Interior News Release,* August 10, 1972.

Valverde, O. and C. V. Dias (1967) *A Rodovia Belém-Brasília.* Fundação IBGE, Inst. bras. Geogr. Série A. [Biblioteca Geográfica Brasileira] Publ. Nº 22. Rio de Janeiro.

Vareschi, V. (1963) La Bifurcación del Orinoco – Observaciones Hidrográficas y Ecológicas de la expedición conmemorativa de Humboldt del año 1958. *Acta Cient. Venez.* 14(4): 98–106.

Varese, S. (1972) Inter-Ethnic Relations in the Selva of Peru. *In* Dostal, W. (ed.) *The Situation of the Indian in South America.* (Symposium on Inter-Ethnic Conflict in South America . . . Bridgetown, Barbados, 25–30 Jan. 1971). Department of Ethnology, University of Berne/World Council of Churches, Geneva.

Veen, A. W. L. (1970) *On geogenesis and pedogenesis in the old coastal plain of Surinam (South America).* Publicaties van het Fysisch-geografisch en Bodemkundig Laboratorium van de Universiteit van Amsterdam Nr. 14. Amsterdam.

Velarde, J. F. (1886) O Rio Madeira e seus affluentes; as ultimas explorações nos rios Beni, Madre de Dios, Orton e Abuná. *Rev. Soc. Geogr.* Rio de. J. 2(3): 165–190.

Venezuela Proposes Vast Waterway System (1972–73) *Venezuela Up-to-Date* (Washington, D. C.) 14(1): 3.

Verissimo, J. (1895) *A pesca na Amazonia.* Livraria Classica de Alves & C., Rio de Janeiro, São Paulo. Republished in 1970 by Universidade do Pará, Belém.

Villas Boas, O. and C. (1968) Saving Brazil's Stone-Age Tribes from Extinction. *Nat. Geogr. Mag.* 134(3): 424–444.

-- (1973) *Xingu, the Indians, their Myths.* Farrar, Straus and Giroux, New York.

-- (1974) Enchentes e o Mato invadem a Estrada de Xavantina-Cachimbo. Statements reported in *O Liberal* (Belém) April 29, 1974.

Wallace, A. R. (1853) *A Narrative of Travels on the Amazon and Rio Negro* . . . Reeve and Co., London.

West India Company (1714) Secret Letter of the Honorable Directors of the Chartered West India Company in Secret Committee, dated May 1, 1714, to Peter van der Heyden Rezen, Commandeur of Essequebo, with directions to explore a river running from the Essequebo to the lake of Parima or Rupowini with a view to taking possession. English Translation *in* Venezuela Nº 3 (1896) Further Documents relating to the Question of Boundary between British Guiana and Venezuela, Annex Nº 43. British Sessional Papers, House of Commons 1896 vol. 97. Command Paper 8106 London H. M. S. O., C–8106.

Williams, C. M., F. C. Kafatos and D. Prescott (1967) Insecticidal Properties of the Waters of the Rio Negro. *R/V 'Alpha-Helix' Amazon Expedition February to October, 1967.* U. S. National Science Foundation/Scripps Institution of Oceanography, University of California, San Diego.

Williamson, A. J. (1923) *English Colonies in Guyana and the Amazon – 1604–1668.* The Clarendon Press, Oxford.

Wijmstra, T. A. and T. v. d. Hammen (1966) Palynological Data on the History of Tropical Savannas in Northern South America. *Leidse Geologische Mededelingen* 38: 71–90.

Xingu tem mais cem hospedes (1974). *O Liberal* (Belém) July 13, 1974.